Augmented Vision and Reality

Volume 10

Series editors

Riad I. Hammoud, Burlington, MA, USA
Lawrence B. Wolff, New York, NY, USA

More information about this series at http://www.springer.com/series/8612

Surya Prakash · Phalguni Gupta

Ear Biometrics in 2D and 3D

Localization and Recognition

 Springer

Surya Prakash
Computer Science and Engineering
Indian Institute of Technology Indore
Indore
India

Phalguni Gupta
Computer Science and Engineering
Indian Institute of Technology Kanpur
Kanpur
India

ISSN 2190-5916   ISSN 2190-5924   (electronic)
Augmented Vision and Reality
ISBN 978-981-287-374-3   ISBN 978-981-287-375-0   (eBook)
DOI 10.1007/978-981-287-375-0

Library of Congress Control Number: 2015933821

Springer Singapore Heidelberg New York Dordrecht London

Printed on acid-free paper

Springer Science+Business Media Singapore Pte Ltd. is part of Springer Science+Business Media
(www.springer.com)

Dedicated to

My Beloved Parents

Surya Prakash

My Wife and Daughter

Phalguni Gupta

Preface

Authentication of a person to ascertain his/her identity is an important problem in society. There are three common ways to perform authentication. The first relies on what a person possesses such as keys, identity cards, etc., while the second is based on what a person knows such as passwords, personal identification numbers (PINs), etc. The third way of authentication relies on what a person carries, i.e., the unique characteristics of a human being (Biometrics). Even though the first two methods are well established and accepted in society, they may fail to make true authentication on many occasions. For example, there is a possibility that items under possession may be lost, misplaced, or stolen. Similarly, one can forget passwords, etc. As a result, authentication may not be correct. However, this is not true in case of biometrics. Thus, most of the limitations of the traditional ways of authentication, which are based on possession and knowledge, can be overcome by the use of biometrics. Since it uses the characteristics of a person's own body or behavior which he/she always carries, there is no chance of forgetting or losing it. Moreover, body characteristics used for authentication are much more complicated and difficult to forge compared to remembering a string (such as password), even of a very long size. The main motivation behind the use of biometrics is to provide a convenient mechanism for person authentication with the help of his/her biological or behavioral characteristics and to eliminate the use of much inconvenient ways of authentication such as that based on ID card, password, physical keys, PINs etc.

There are two types of characteristics used in biometrics for person authentication. The first type of characteristics are physiological in nature, while the other is based on the behavior of human beings. Physiological characteristics depend on "what we have" and derives from the structural information of the human body, whereas behavioral characteristics are based on "what we do" and depend on the behavior of a person. The unique biometric characteristic (be it physiological or behavioral) which is used for authentication is commonly referred as a biometric trait. Common examples of physiological biometric traits are face, ear, iris, fingerprint, hand geometry, hand vein pattern, palm print, etc., whereas signature, gait (walking pattern), speech, key strokes dynamics, etc., are examples of behavioral biometrics.

Among various physiological biometric traits, the ear has gained much popularity in recent years as it has been found to be a reliable biometrics for human recognition. The use of ear for human recognition has been studied by Iannarelli in 1989. This study suggested the use of features based on 12 manually measured distances of the ear. It has used 10,000 ear images to demonstrate the uniqueness of ears and has concluded that ears are distinguishable based on a limited number of characteristics. This has motivated researchers in the field of biometrics to look at the use of ear for human recognition. Analysis of the decidability index (which measures the separation between genuine and imposter scores for a biometric system) also suggests the uniqueness of an individual ear. It has been found that the decidability index of the ear is in an order of magnitude greater than that of face, but not as large as that of iris. Below is a list of characteristics which make ear biometrics a popular choice for human recognition.

1. Ear is found to be very stable. Medical studies have shown that major changes in the ear shape happen only before the age of 8 years and after that of 70 years. The shape of the ear is found to be stable for the rest of life.
2. Ear is remarkably consistent and does not change its shape under expressions like face.
3. Color distribution of the ear is almost uniform.
4. Handling background in case of ear is easy as it is very much predictable. An ear always remains fixed at the middle of the profile face.
5. Ear is unaffected by cosmetics and eye glasses.
6. Ear is a good example of passive biometrics and does not need much cooperation from the subject. Ear data can be captured even without the knowledge of the subject from a distance.
7. Ear can be used in a standalone fashion for recognition or it can be integrated with the face for enhanced recognition.

Ear recognition consists of two important steps and they are (i) Ear detection and (ii) Recognition. Ear detection carries out the segmentation of the ear from profile face before using it for recognition task. Most of the ear recognition techniques directly work on manually segmented ear images, however, there also exist a few approaches which can take complete side face image as input and segment the ear automatically. This book will present some efficient but automatic ear detection techniques for 2D as well as for 3D.

Recognition step deals with the task of human recognition based on the segmented ear. Major challenges in 2D ear recognition are due to poor contrast and illumination, presence of noise in the ear image, poor registration of gallery (database), and probe images. Challenges in 3D ear recognition arise mainly from poor registration of gallery and probe images and presence of noise in the 3D data. This book presents efficient recognition techniques both in 2D and 3D, which have attempted to overcome these challenges.

This book consists of five chapters. A brief description of the content of each chapter is as follows. Chapter 1 is an introductory chapter and presents the basics of a biometric system, different biometric traits, various performance measures, information about various publicly available ear databases, etc.

Chapter 2 talks about ear detection in 2D. It first reviews existing ear localization techniques available in the literature and subsequently presents an efficient recently proposed ear localization technique in detail. This ear localization technique is found to be invariant to scale, rotation, and shape. It makes use of connected components of a graph constructed with the help of edge map of the profile face image to generate a set of probable ear candidates. True ear in this technique is detected by performing ear identification using a rotation, scale, and shape invariant ear template.

Chapter 3 talks about ear recognition in 2D. It starts with presenting a review on ear recognition techniques available in the literature. Further, in detail it describes a recently proposed efficient ear recognition technique in 2D which makes use of multiple image enhancement techniques and local features based on Speeded Up Robust Features (SURF). The use of multiple image enhancement techniques in it has made it possible to counteract the effect of illumination, poor contrast, and noise while SURF-based local feature helps in matching the images that are not properly registered and suffer from pose variations. For a given ear image, this technique obtains three enhanced images which are used by SURF feature extractor to generate three sets of SURF features for an ear image. Three nearest neighbor classifiers are, respectively, trained on these three sets of features and finally results in all the classifiers fused to get the final result.

Chapter 4 starts with presenting a review on 3D ear detection technique and subsequently discusses in detail a recently proposed technique for ear detection in 3D. As we know, detection of ears from an arbitrary 3D profile face range image is a challenging problem due to the fact that ear images can vary in scale and pose under different viewing conditions. The technique discussed in this chapter is capable of handling these issues due to variations in scale and rotation. Moreover, this technique does not require any registered 2D image for ear detection in 3D. Also, it can detect left and right ear at the same time without imposing any additional computational cost.

Chapter 5 focuses on ear recognition in 3D. It first presents a detailed review of human recognition techniques in 3D. Further, it discusses a recent human recognition technique which makes use of 3D ear data along with registered 2D ear images. The technique first coarsely aligns the 3D ear data using local features computed from registered 2D ear images and then uses Generalized Procrustes Analysis and Iterative Closest Point (GPA-ICP)-based matching technique for final alignment. It integrates GPA with ICP to achieve robust 3D ear matching. Coarse alignment of the data before applying GPA-ICP helps to provide a good initial point for GPA-ICP-based matching algorithm.

Contents

Abbreviations

ADHist	Contrast Limited Adaptive Histogram Equalization
AUC	Area Under ROC Curve
CMC	Cumulative Matching Characteristics
CRR	Correct Recognition Rate
EER	Equal Error Rate
EOP	Extended Orthogonal Procrustes
EUC	Error Under ROC Curve
FAR	False Acceptance Rate
FMR	False Match Rate
FNMR	False Non-Match Rate
FRR	False Rejection Rate
GAR	Genuine Acceptance Rate
GLOH	Gradient Location and Orientation Histogram
GPA	Generalized Procrustes Analysis
GPA-ICP	GPA Integrated ICP Technique
HCS	Histograms of Categorized Shapes
ICA	Independent Component Analysis
ICP	Iterative Closest Point Technique
IITK	Indian Institute of Technology Kanpur
LFGPA-ICP	Local 2D Features and GPA based Improved ICP Matching Technique
LSP	Local Surface Patch
NLM	Non-Local Mean Filter
PCA	Principal Component Analysis
RBF	Radial Basis Function
ROC	Receiver Operating Characteristics
ROI	Region of Interest
SF	Steerable Filter
SFFS	Sequential Forward Floating Selection
SIFT	Scale Invariant Feature Transform
SNR	Signal-to-Noise Ratio

SURF	Speeded Up Robust Features
UCR	University of California Riverside
UND	University of Notre Dame
WEOP	Weighted Extended Orthogonal Procrustes

Chapter 1
Introduction

Authentication of a person's identity is a very old but challenging problem. There are three common ways which are used for authentication. First one is based on what a person has (Possession) such as keys, identity cards etc. Second mode of authentication is based on what a person knows or remembers (Knowledge) such as passwords, PINs etc. Third way of authentication is based on what a person carries, i.e. the characteristics of a human being (Biometrics). There are chances that the items which are under possessions may be lost and knowledge may be forgotten. But this is not the case with Biometrics. Limitations of the first two methods can be overcome if ones makes use of particular characteristics of the body or habits as the mode of authentication because they are difficult to forget or forge. This is the main driving force behind biometrics based authentication getting more and more popularity day-by-day. The purpose of using a biometrics is to provide a mechanism to recognize a person with the help of his/her biological characteristics and to eliminate the use of much inconvenient ways of recognition which are based on ID card, password, physical keys etc.

The term "Biometrics" is associated with the use of certain physiological or behavioral characteristics to authenticate or identify an individual. It can be defined as a measurable characteristic of a person which can be used for automatically recognizing his/her identity. Physiological characteristics are based on "what a person has" and rely on the structural information of human body whereas behavioral characteristics are based on "what a person does" and are dependent on the behavior of a person.

The idea behind biometrics based recognition is to use these special unique characteristics of a person available in face, ear, iris, fingerprint, signature etc. It is evident that use of biometrics adds a complexity to the authentication system that would be hard to reach with a standard password-based technique. Common reasons for which method of authentication using biometrics is preferred over traditional passwords and PIN based methods are discussed below.

© Springer Science+Business Media Singapore 2015
S. Prakash and P. Gupta, *Ear Biometrics in 2D and 3D*,
Augmented Vision and Reality 10, DOI 10.1007/978-981-287-375-0_1

1. **Physical presence**: The person to be identified is required to be physically present at the time of authentication in biometric based authentication. This makes biometrics based authentication secure.
2. **No need for remembering information**: Authentication based on biometric techniques obviates the need to remember a password or a PIN. Information used in biometric authentication is always carried by the person with him/her.
3. **Less prone to forgery**: There is less possibility of biometric identity to be faked, forged and fooled.

Moreover, biometric systems can be used in conjunction with passwords and PINs, thus improving the security of existing systems without replacing them.

1.1 Biometric Properties

Clarke [1] has analyzed the requirements of a biometric system and has suggested the following properties that a biometric characteristic/trait should possess to make itself suitable for successful authentication.

1. **Universality**: Every person should have the biometric characteristic and it should seldom lose to an accident or disease.
2. **Uniqueness**: No two persons should have the same value of the biometric characteristic i.e. it should be distinct across individuals.
3. **Permanence**: Biometric characteristic should not change with time. It should not subject to considerable changes based on age or disease.
4. **Collectability**: Biometric characteristic should be collectable from anyone on any occasion.
5. **Acceptability**: Society and general public should have no objection to provide the biometric characteristic.
6. **Measurability**: Measurability is meant for the possibility of acquiring and digitizing the biometric characteristic using some suitable digital devices/sensors without causing any inconvenience to the person.
7. **Circumvention**: A biometric characteristic can be imitated or forged. By circumvention it is meant that the system should be able to handle these situations effectively.

Unfortunately, it is very hard to have a biometrics satisfying all the above issues fully. Depending upon the needs of the application, one should select the most appropriate biometrics. Table 1.1 compares some of the biometric traits from the point of view of these properties.

Table 1.1 Comparison of some of the biometric traits from the point of view of properties discussed in Sect. 1.1

Trait	Universality	Uniqueness	Permanence	Collectability	Acceptability	Measurability	Circumvention
Face	High	High	M	High	High	High	Low
Fingerprint	Medium	High	Medium	Medium	Medium	High	High
Ear	Medium	High	High	High	High	High	Low
Iris	Medium	High	Medium	Low	Low	High	Low
Palm print	Medium	Medium	Medium	Medium	Medium	High	Medium
Signature	Low	High	Medium	Medium	Medium	High	High
Voice	High	Medium	Medium	High	High	High	High
Gait	High	Medium	Medium	High	High	Medium	Low
Keystrokes	Medium	Medium	Medium	Medium	High	Medium	Medium

1.2 Operational Modes of a Biometric System

A biometric system can use either one trait or multiple traits for person authentication.
Systems using one trait are called unimodal biometric systems while those using more
than one trait are called multi-modal biometric systems.

Any biometric system can be used in three modes: (i) as an enrollment system,
(ii) as a verification system and (iii) as an identification system. Enrollment is used
to register a new person with the system. Verification involves validating a claimed
identity of a person and involves one to one comparison. Identification is a process
of searching identities in the database containing biometric templates. These modes
of operations are explained below in detail.

1.2.1 Enrollment

It is the first step to use any biometric system. It deals with the enrollment of subjects
in the system. Enrollment process consists of three major steps *viz.* data acquisition,
preprocessing and feature extraction. Data acquisition deals with the collection of
raw data from subjects for specific trait(s). Preprocessing step performs data cleaning
and noise removal in the collected data. It also detect the Region of Interest (ROI) in
the acquired image. Feature extraction process extract features from the ROI region.
Enrollment process is completed by registering (storing) these features in the database
against a unique ID.

Figure 1.1 shows the block diagram of a biometric system in enrollment mode.
Once the enrollment is over, biometric system can be used for authentication which
is usually carried out in two modes: verification and identification. These two modes
are explained in subsequent sections.

1.2.2 Verification

Biometric verification is a way by which a person can be uniquely identified by
evaluating one or more distinguishing biometric characteristics (also called features

Fig. 1.1 Block diagram of
enrollment module

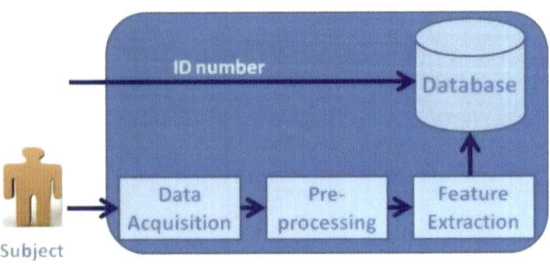

Fig. 1.2 Block diagram of a
verification system

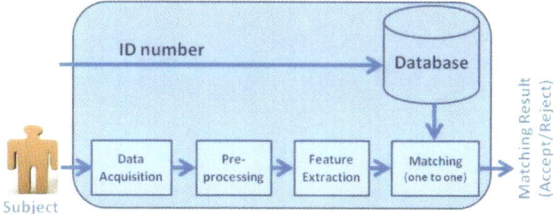

or traits). In a verification system one has to prove his/her identity to get access of the system. It takes a unique identity number and biometric characteristics of a person as input and compares with the enrolled characteristics in the database against the input identity number. Depending upon the matching score, verification system accepts or rejects the claimed identity.

A verification system consists of four subsystems *viz.* Data acquisition module, Preprocessor, Feature extractor and Matcher which work in serial fashion. Data acquisition module is responsible to collect raw data from the subject. Pre-processor is used for removing noise and locating the ROI in the input image. Feature extractor computes the salient features of the input biometric data. Matcher is used to compare feature vectors of two biometric identities and it yields a similarity score that indicates how closely the two compared feature vectors are related. Higher similarity score produced by a matcher indicates better matching. If the score is sufficient enough to declare a match, the claimed identity is accepted; otherwise it is rejected. The rules governing the declaration of a match are often configured by the end user so that he/she can determine how the biometric system should behave based on security and operational considerations. Verification process uses one-to-one comparison which means input biometric features for a person are only compared with the one registered against the claimed identity.

Figure 1.2 shows the block diagram of a verification system. Output of a verification system is either accept or reject. If the input biometric features for a person matches with the one already enrolled in the database, he/she is accepted by the system; otherwise rejected. A common example of a verification system is a biometrics enabled ATM machine, where one has to swap the card to tell his/her identity number and to produce biometric features to the machine for comparison to withdraw money.

1.2.3 Identification

Identification involves searching of subjects similar to the query template present in the database. Like a verification system, an identification system also consists of four subparts *viz.* Data Acquisition module, Pre-processor, Feature extractor and Matcher. These modules function in the same way as in case of verification except the slight modification in matching module. In verification, comparison is performed as one-to-one whereas in identification, all stored templates in the database are compared

Fig. 1.3 Block diagram of
an identification system

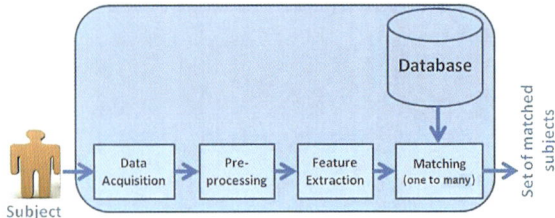

(searched) to produce best possible matches of a query template. Due to this, process
of identification requires one-to-many comparison. Output of an identification system
is a set of subjects which are the best possible matches to the query template. Usually
in these systems, one is interested to find out the top m similar subjects from the
database of size n where $m \leq n$.

An important difference between identification and verification is that unlike
verification, identification does not require a claimed identity and performs matching operation against the entire database. Figure 1.3 shows the block diagram of an
identification system.

Positive and Negative Identification: Based on the user cooperation during the
identification process, it can be classified into two types: positive identification and
negative identification. In positive identification, a subject is interested to be identified
by the system. A good example of positive identification can be seen when a person
tries to get access to some restricted area using his biometric feature such as face or
fingerprint.

In negative identification, a subject tries to avoid his/her successful identification
by the system. In this case, subject is non-cooperative and does not want to be
identified. He/she may not like to cooperate in providing the biometric data and hence
often supervision is required at the time of data acquisition and feature extraction.
An example of a negative identification can be seen in a thief interested in not being
identified by the system with the help of latent fingerprints found at the scene of
crime.

Depending upon the type of identification, it may be required to use different
kind of sensors in data acquisition. For example, negative identification may need
more data for identification. Hence, a fingerprint based system may need full size
sensors and ten-print treatment of the fingerprint data at the time of enrolment and
identification to get desirable results in case of negative identification.

1.3 Performance Measures

Performance of a biometric system deals with the quantifiable assessment of the accuracy and other characteristics of the system. Performance of a biometrics system can
be measured for three tasks: ROI detection, verification and identification. Following

are few important metrics which are commonly used to evaluate the performance of a biometric system.

1.3.1 ROI Detection

Performance of ROI detection (ear detection) can be measured as follows.

$$\text{Detection Accuracy} = \frac{\text{Number of Correct Detections} \times 100}{\text{Total Test Samples}} \% \qquad (1.1)$$

1.3.2 Verification Accuracy

It is measured using following parameters.

1. **False Acceptance Rate** (*FAR*): It is defined as the fraction of candidates falsely accepted by a biometric system. That means, it is the rate at which an imposter is incorrectly accepted as genuine person. A false acceptance may lead to damages and it occurs when matching score established by a biometric system for an imposter satisfies the threshold criteria of matching. Low value of *FAR* shows that the biometric system can efficiently capture the inter-class variability through its feature representation and matching. *FAR* which is also sometime referred as False Match Rate (*FMR*), is given by

$$FAR = \frac{\text{Number of Imposters Accepted} \times 100}{\text{Total Number of Imposter Comparisons}} \% \qquad (1.2)$$

2. **False Rejection Rate** (*FRR*): It represents the fraction of candidates falsely rejected by a biometric system. In other words, it is the rate at which a genuine person is incorrectly rejected as an imposter. *FRR* is also called as False Non-Match Rate (*FNMR*). Low value of *FRR* shows that the system can capture intra-class variations efficiently through its feature representation technique and matching. Thus, *FRR* is given by

$$FRR = \frac{\text{Number of Genuine Persons Rejected} \times 100}{\text{Total Number of Genuine Comparisons}} \% \qquad (1.3)$$

Genuine Acceptance Rate (*GAR*) measures the faction of the acceptance of genuine candidates and is defined by

$$GAR = (100 - FRR) \% \qquad (1.4)$$

Figure 1.4 shows an example of *Threshold* versus *FAR* and *FRR* curves.

Fig. 1.4 *Threshold* versus *FAR*, *FRR* curves

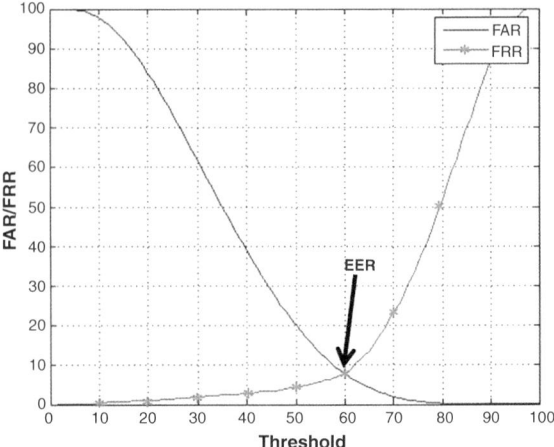

3. **Equal Error Rate** (*EER*): It is defined as the rate at which both *FAR* and *FRR* errors are equal, i.e.

$$EER = FAR \text{ for which } FAR = FRR \qquad (1.5)$$

4. **Recognition Accuracy**: It is used to measure the performance of a verification system and is defined as

$$\text{Recognition Accuracy} = \left(100 - \frac{FAR + FRR}{2}\right)\% \qquad (1.6)$$

Threshold used for matching plays an important role in deciding the optimum values of *FAR* and *FRR*. Any change in the threshold value makes the changes in *FAR* and *FRR*. A combination of optimum *FAR* and *FRR* is chosen to define the accuracy. Often, the combination of *FAR* and *FRR* which gives the highest accuracy is considered as the optimum.

5. **Receiver Operating Characteristics** (*ROC*) **Curve**: The performance of a verification system can also be evaluated using a *ROC* curve. It graphically demonstrates the changes of *GAR* (Genuine Acceptance Rate) with respect to changes in *FAR*. It measures the ability of a system to discriminate genuine persons from imposters. Since *ROC* curve plots the graph between *FAR* and *GAR* (or *FRR*) values and hence eliminates the use of threshold parameter in the graph. An ideal *ROC* curve would include a point at *GAR* = 100, *FAR* = 0. However, in a real scenario, it is difficult for a biometric system to achieve such perfect performance. Figure 1.5 shows an example of *ROC* curve. *ROC* curve provides a good way to compare the performance of two biometric systems.

Fig. 1.5 Receiver operating characteristic (*ROC*) curve

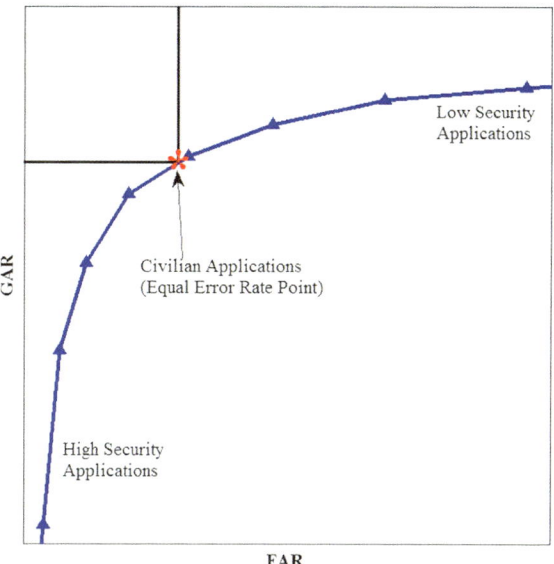

6. **Error Under *ROC* Curve (*EUC*)**: Area under the *ROC* curve (*AUC*) is defined as a scalar quantity which tells the probability that a classifier gives a higher match score to a randomly selected genuine sample than to a randomly selected impostor sample. Commonly, for a better interpretation, the Error under the *ROC* Curve (*EUC*) is used and is defined as follows.

$$EUC = (100 - AUC)\,\% \tag{1.7}$$

1.3.3 Identification Accuracy

It is usually measured in terms of Cumulative Matching Characteristics and Correct Recognition Rate. These terms can be explained as follows.

1. **Cumulative Matching Characteristics (*CMC*)**: It is a measure to evaluate the performance of a biometric identification system and is computed by comparing the database templates with the test templates and ranking them based on their matching scores. It is also called Rank-k recognition rate. It shows how often the genuine subject's template appears within rank-k (for example 1, 5, 10, 100 etc.) matches based on the matching score. It can be defined as follows.

$$CMC = \frac{\text{Number of Genuine Matches that Occurred in Top } k \text{ Matches}}{\text{Total Number of Test Matches Performed}} \times 100\,\% \tag{1.8}$$

2. **Correct Recognition Rate** (*CRR*): It is the most commonly used performance
   measure to evaluate a biometric identification system. It is also computed by comparing the database templates with the test templates and ranking them based on
   their matching scores. It shows how often the genuine subject's template appears
   in rank-*1* match based on the matching score. It is also called Rank-*1* recognition
   rate and can be defined as follows.

$$CRR = \frac{\text{Number of Genuine Matches that Occurred as the Top-}1\text{ Match}}{\text{Total Number of Test Matches Performed}} \times 100\,\%$$

$$(1.9)$$

1.4 Biometric Traits

Biometric characteristics may be of either physiological or behavioral in nature.
Few well known physiological and behavioral biometric traits are discussed in this
section.

1.4.1 Physiological Traits

This subsection provides basic information about few popular physiological biometric namely face, fingerprint, ear, iris and palm print. Examples of these traits are
shown in Fig. 1.6.

- **Face**: Although the concept of recognizing someone from facial features is
  intuitive, facial recognition makes human recognition a more automated process.
  A typical face recognition system starts with detection of the location of the face in
  the acquired image. From the detected face, distinguishing features are extracted.
  Features of a test image are compared with those of images stored in the database
  and decision is made based on the matching score against a threshold. What sets
  apart facial recognition from most of other biometrics is that it can be used for
  surveillance purposes. For example, public safety authorities who want to locate
  certain individuals such as wanted persons can use facial recognition systems.
  Since faces can be captured from a distance, facial recognition can be done without any physical contact. This feature gives facial recognition a covert capability.
  Though face recognition has been used for long time in human recognition, it is
  still challenging to get an efficient system due to several reasons. A face image gets
  changed due to different illumination, facial expressions and aging. It may also get
  changed due to different pose (orientation). Occlusion also imposes challenges in
  development of a robust face recognition system.

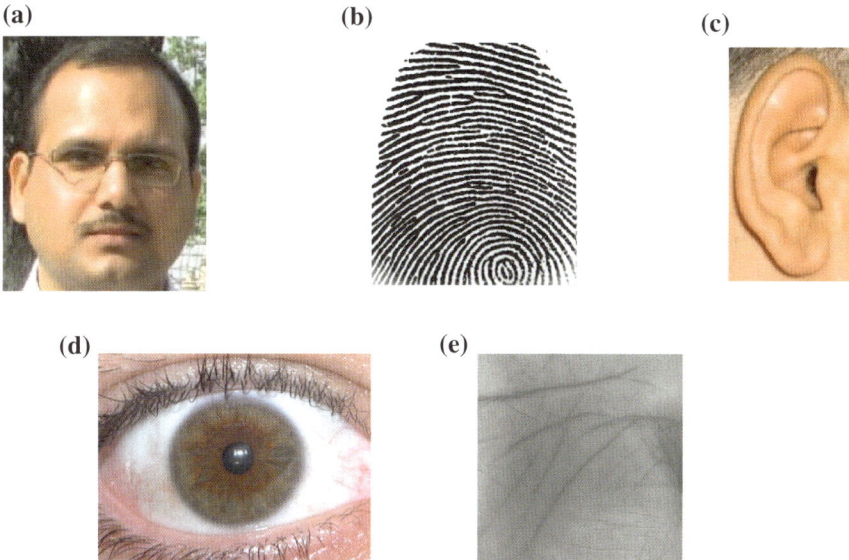

Fig. 1.6 Examples of physiological biometric traits. **a** Face. **b** Fingerprint. **c** Ear. **d** Iris. **e** Palm print

- **Fingerprint**: Fingerprint has been used for person recognition for many centuries and is the most frequently used biometrics. A fingerprint pattern is composed of ridges and valleys. Ridges present various kinds of discontinuities which provide invariant and discriminatory information of a person. These points of discontinuities are called minutiae points and are used to recognize a person. Depending upon the structure, minutiae points can be classified into many types such as ridge bifurcation, ridge ending, isolated points etc. In a fingerprint recognition system, ridge bifurcations and ridge endings are commonly used. Fingerprint images can be extracted either by creating an inked impression of the fingertip on paper and then digitizing it or by directly using digital sensors. Captured images are further used by the feature extraction module to compute the feature values (typically the position and orientation of minutiae points). Fingerprint matching determines the degree of similarity between two fingerprints by comparing their ridge structures and/or the spatial distribution of the minutiae points. Fingerprint is one of the most economical, developed and standardized biometrics. Its feature template is small in size; hence it needs a less storage space and less matching time. Though fingerprint biometrics has come a long way, it still faces few challenges. It is very much prone to spoofing. It is difficult to perform fingerprint recognition for wet or dirty fingers. Quality of the fingerprint also imposes big challenge in achieving good performance. For example, there may be several people such as laborers, farmers, field workers etc. who may not have good quality fingerprint data.

Acceptability is also an issue as there are still few people who get offended in giving there fingerprint data due to perception of its relation to criminal identification.

- **Ear**: Although relatively a new biometrics, ear has been used as a means of human recognition in forensic field for long time. During crime scene investigation, ear-marks[1] and earprints[2] have often been used for recognition of a suspect in the absence of valid fingerprints. Similar to fingerprints, the long-held history of the use of earmarks suggests its use for automatic human recognition. An ear recognition system is very much similar to a typical face recognition system. However, ear has few advantages over the face. For example, its appearance does not change due to expression and it is found to be unaffected by aging process. Its color is uniform and background is predictable.

- **Iris**: Flowery textural pattern around the eye ball in the eye is called iris. This textural pattern is found to be unique for everyone and can be used for the recognition of an individual. Iris recognition uses high-resolution image of the iris of an individual and has been found to have the smallest outlier group among all biometric technologies. For iris recognition, an image of human eye is obtained using an infrared (IR) based camera and iris portion is localized. Further, features are extracted using some pattern recognition technique and compared with the stored templates in the database to find a match. A key advantage of iris recognition is its stability and template longevity because apart from trauma, a single enrolment can last for a lifetime. It has also been found that efficiency of iris recognition is not hindered by glasses or contact lenses. Though iris recognition produces a very high accuracy, there are some issues with it. It needs much user cooperation for data acquisition and often sensitive to occlusion. Iris data acquisition is very intrusive and needs a very controlled environment. Also, data acquisition devises are quite costly. Iris recognition cannot be used in a covert situation.

- **Palm print**: Palm print offers a reliable means of recognizing a person due to its stability and uniqueness. A human palm print is made up of principal lines, wrinkles and ridges which are found to be unique for individuals and are used for recognition. Palm print recognition uses many of the same matching characteristics that are used in fingerprint recognition. Compared to fingerprint recognition, palm print recognition offers an advantage. Since area of palm print is larger than fingerprint, there is a possibility of capturing more distinctive features in it. However, there are few disadvantages of palm print recognition over fingerprint recognition as well. Since it captures large area, it needs more processing time. Palm print scanners are even bulkier than fingerprints scanners because they need to capture relatively a larger area.

---

[1] Impression of human ear recovered by some means from a crime scene.
[2] Control impression taken from the ear of an individual.

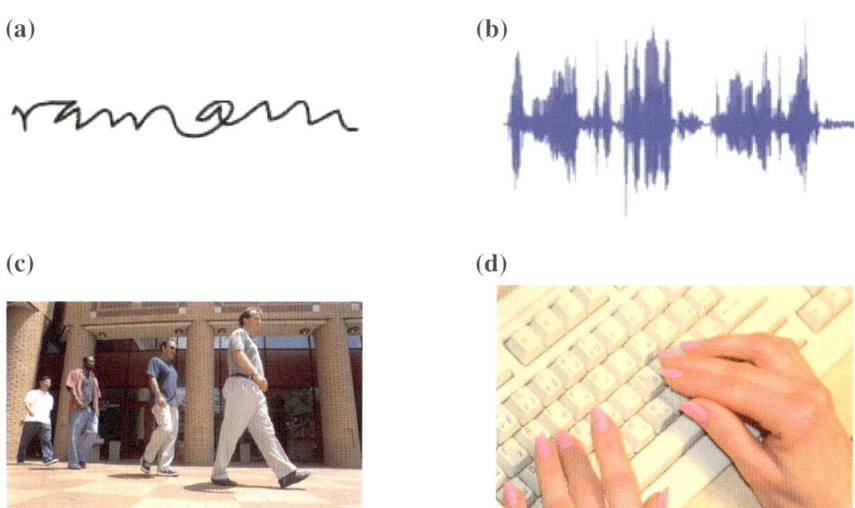

Fig. 1.7 Examples of behavioral biometric traits. **a** Signature. **b** Voice. **c** Gait. **d** Keystroke

1.4.2 Behavioral Traits

This subsection provides the basic information about few popular behavioral biometric traits namely signature, voice, gait and key-stroke dynamics. Examples of these traits are shown in Fig. 1.7.

- **Signature**: It is a handwritten representation of a person's identity. It belongs to the behavioral biometric characteristics of a person as it depends on the way a person puts his/her signature. This type of biometric system records several characteristics of the signing style of a person to carry out the recognition task. Signing style includes the angle of writing, amount of pressure employed in writing, formation of letters, number of key strokes etc. These characteristics are recorded and used in developing a signature based biometric system. A signature recognition system can work in two modes: Off-line mode and On-line mode. In Off-line mode, it uses gray-scale signature images that have been previously captured with the help of a scanner or a camera. In On-line mode, signature is captured online at the time of signing itself on a device such as pen tablets, pen displays, touch screens etc. The advantage with signature recognition is that it employs less expensive hardware as compared to some of the high end biometric systems such as iris recognition etc. Also, verification time in signature recognition is less because it uses a low dimensional feature template. The major challenge faced by this technology is the poor permanence because of the high degree of variability in handwriting with time. An individual's signature can substantially vary over a lifetime. Other challenges include low universality (as everyone may not be able to put signature) and vulnerability to forgeries.

- **Voice**: Every individual has distinct voice characteristics such as different voice texture, unique pronunciation style etc. Voice recognition (also referred as voice biometrics or speaker recognition) is a type of behavioral biometrics which uses voice characteristics of a person to recognize him/her with the help of pre-stored voice templates. It is sometime considered as physiological. This is due to the fact that voice characteristics depend on the shape of the vocal track. Hence, it is the only form of biometrics which uses both behavioral as well as physiological features. Voice biometrics extracts information from the stream of speech and makes use of lots of data, microphones and noise cancelation software to perform its task. Systems using voice biometrics have been applied to many real-world security applications for more than a decade. Its use is increasing rapidly in a broad range of applications such as financial, retail, entertainment, crime investigation etc. Voice recognition is non-intrusive and has high social acceptability. It also offers a cheap recognition technology because general purpose voice recorders can be used to acquire the data. However, a persons voice can be easily recorded and can be used for unauthorized access. An illness such as cold can change a persons voice, making voice recognition difficult.
- **Gait**: Human recognition based on gait is relatively recent as compared to other traditional approaches such as fingerprint and face recognition. Gait is defined as the style of walking. The psychophysical studies have revealed that people can easily identify known individuals based on their gait. This has led to the use of gait as a biometrics for human recognition. Gait based biometric system usually consists of a video camera which can capture images of a person walking within its field of view. Appropriate features such as silhouettes, shape, joint angles, structure and motion are extracted from the video to form gait signature of an individual. A gait system can easily be deceived because walking patterns can sometime be altered. The ability of gait biometrics to perform well in real life scenario such as airports and railway stations is not yet proven.
- **Keystrokes Dynamics**: Keystroke dynamics (also called typing dynamics) refers to the detailed timing information which tells exactly the time of pressing each key and that of releasing while typing on computer keyboard by a person. The manner in which a person types on a computer keyboard is distinguishable from another person and is found to be unique enough to determine the identity of a person. Keystroke biometrics is a type of behavioral biometrics and is based on keystroke dynamics. The unique behavioral characteristics which are used in keystroke biometrics include typing speed, key holding time, time elapsed between two consecutive keystrokes, the sequence of keys used to type an uppercase letter etc. Typing pattern in keystroke biometrics is usually extracted from computer keyboards; however this information can also be extracted from any input device such as mobile phones, palm tops etc. having keys. Keystroke recognition systems face almost similar type of problems that one faces with a username/password based system. For example, passwords can be forgotten and an individual has to remember multiple passwords to gain access to different systems. Also keystroke recognition is not yet proven to be unique for all individuals and has not yet been tested on a large scale.

1.5 Motivation

Among various physiological biometric traits, ear has received much attention in recent years as it has been found to be a reliable biometrics for human recognition [2]. A very early study on use of ear for human recognition has been carried out by Iannarelli in [3]. This study has proposed a manual ear based recognition system which has used twelve features of the ear. These features represent manually measured distances between specific ear feature points. The system has used 10,000 ear images to find the uniqueness criteria between any two ears. This study has suggested that ears may be distinguishable based on limited number of characteristics and features which has motivated researchers to use ear for human recognition. Analysis of the decidability index (which represents the separation between genuine and imposter scores for a biometric system) also indicates the uniqueness of an individual ear where the decidability index of ear is found to be an order of magnitude greater than that of face, but not as large as that of iris. The characteristics making ear biometrics much popular are given below.

1. Ear is remarkably consistent and does not change its shape under expressions like face. Moreover, ear has uniform color distribution.
2. Changes in the ear shape happen only before the age of 8 years and after that of 70 years [3]. Shape of the ear is very much stable for the rest of the life.
3. In face, handling background is a challenging issue and often it requires data to be captured under controlled environment. However, in case of ear, background is predictable as an ear always remains fixed at the middle of the profile face.
4. Size of the ear is larger than fingerprint, iris, retina etc. and smaller than face, and hence ear can be acquired easily.
5. Ear is not affected by cosmetics and eye glasses.
6. Ear is a good example of passive biometrics and does not need much cooperation from user. Ear data can be captured even without the knowledge of the users from a far distance.
7. Ear can be used in a stand alone fashion for recognition or it can be integrated with the face for enhanced recognition.

An ear recognition consists of two major steps and they are (i) Ear detection and (ii) Recognition. Ear detection deals with the segmentation of ear from profile face before using it for recognition task. Most of the well known recognition techniques work upon manually segmented ear images. There exists few ear recognition techniques which are capable of taking complete face image as input has mechanism to detect and crop ear for recognition. In this book, we present efficient but automatic ear detection techniques for 2D as well as for 3D.

Recognition deals with the recognition of a person based on the segmented ear. Major challenges in 2D ear recognition come from poor contrast and illumination, presence of noise in the ear image, poor registration of database and probe image. Challenges in 3D ear recognition arise mainly from poor registration of database and probe image and presence of noise in the 3D data. We shall present efficient recognition techniques both in 2D and 3D which have attempted to overcome these challenges.

1.6 Popular Ear Databases

This section discusses three popular ear databases, namely IIT Kanpur (IITK) database, University of Notre Dame-Collection E (UND-E) [4] and University of Notre Dame-Collection J2 (UND-J2) [5]. Following subsections present details of these databases.

1.6.1 IITK Database

It consists of three data sets. Data Set 1 contains 801 profile face images collected from 190 subjects, 2 or more images per subject. These images include frontal view of the ears. Few sample images from Data Set 1 are shown in Fig. 1.8.

Data Set 2, whose acquisition setup is shown in Fig. 1.11a, contains 801 profile face images collected from 89 subjects, 9 images per subject for various in-plane rotations and scales. Images contain frontal view of the ear taken at three different positions, a person looking straight, the person looking at 20° (approx) down and looking at 20° (approx) up. At all these 3 positions, images are captured at 3 different scales by setting the digital zoom of the camera at 1.7x, 2.6x and 3.3x and positioning the camera at a distance of about 1 m. Figure 1.9 shows 9 images for an individual.

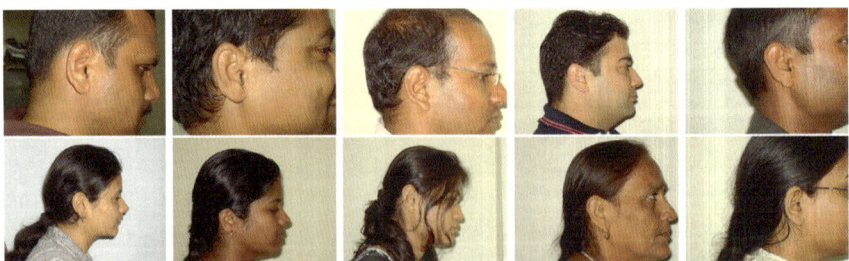

Fig. 1.8 Sample images from IITK Data Set 1

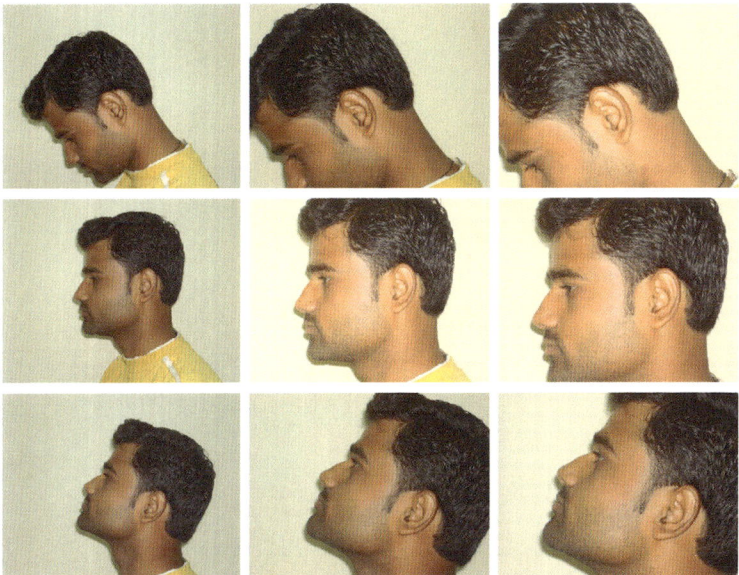

Fig. 1.9 Sample images from IITK Data Set 2

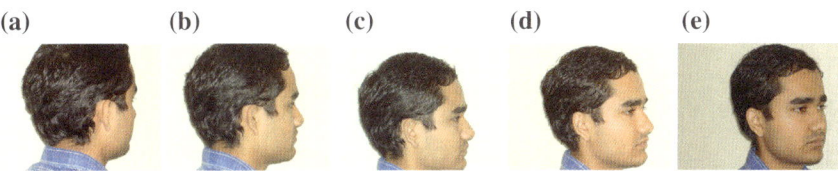

Fig. 1.10 Sample posed images for an individual from IITK data set 3. **a** −40°. **b** −20°. **c** 0°. **d** 20°. **e** 40°

Data Set 3 consists of complex images captured for various out-of-plane rotations from 107 subjects. The camera is moved on a circle with the subject assumed to be at the center of the circle. Camera facing the frontal of the ear is considered as 0°. Profile face images are captured at −40°, −20°, 0°, +20° and +40° placing the camera tripod at fixed landmark positions. Two images for each pose (angle) are obtained, producing 10 images per subject. The acquisition setup used to acquire this data set is shown in Fig. 1.11b. This data set contains 1070 images collected from 107 subjects. Figure 1.10 shows a sample snapshot of angular posed profile face images of a subject.

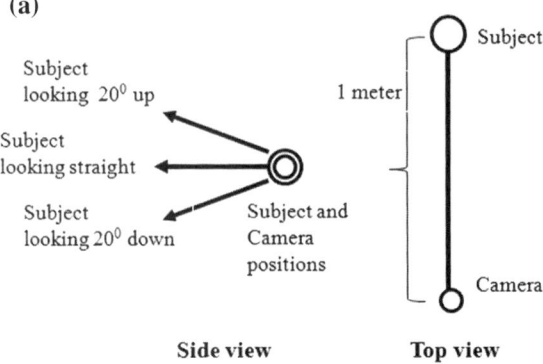

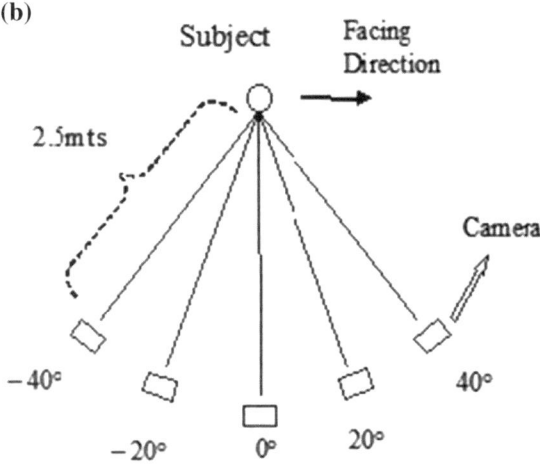

Fig. 1.11 Data acquisition setup used in collection of IITK database. **a** Data Set 2. **b** Data Set 3

1.6.2 University of Notre Dame-Collection E (UND-E) Database [4]

This database consists of 464 profile face images collected from 114 subjects, 3 to 9 samples per subject. Images are collected on different days with different conditions of pose and illumination. It can be noted that there exist a huge intra-class variation in these images due to pose variation and different imaging conditions. Few sample images of this database are shown in Fig. 1.12a.

(a)

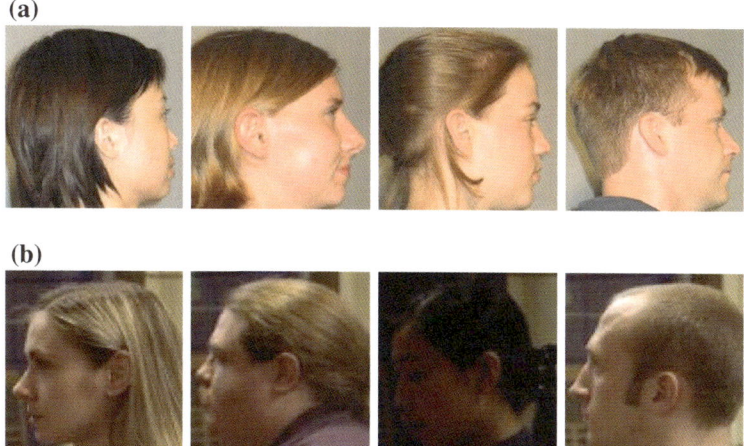

(b)

Fig. 1.12 Sample 2D images from UND database. **a** UND-E Data Set. **b** UND-J2 Data Set

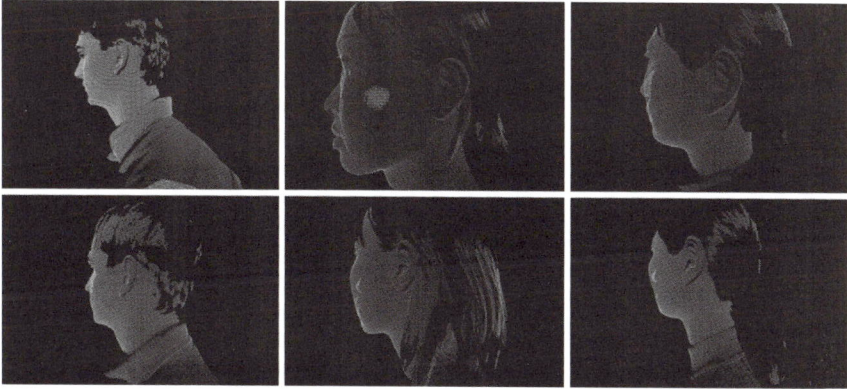

Fig. 1.13 Sample 3D profile face range images for three subjects (respective columns) with pose and scale variations from UND-J2 database

1.6.3 University of Notre Dame-Collection J2 (UND-J2) Database [5]

This database consists of 2414 2D profile face images (with some duplicates) along with registered 3D range images obtained from 404 subjects. After removal of duplicates, there exist 1780 2D (and corresponding registered 3D) profile face images . Few sample 2D and 3D profile face images from the database are shown in Figs. 1.12b and 1.13 respectively. Table 1.2 provides summary of all these databases.

Table 1.2 Summary of the databases used in experimentation

Database	Number of subjects	Total samples	Description
IITK Data Set 1	190	801	2–10 2D images per subject, frontal ear images
IITK Data Set 2	89	801	9 2D images per subject, frontal ear images affected by scaling, rotation and poor registration
UND Dataset (Collection E)	114	464	3–9 2D images per subject, images affected by illumination and pose variations, poor contrast and registration
UND Dataset (Collection J2)	404	1780	2–13 3D (with corresponding 2D) images per subject, images affected by illumination and pose variations, poor contrast and registration, images collected in two sessions with a gap of at least 17 weeks

References

1. Clarke, R. 1994. Human identification in information systems: Management challenges and public policy issues. *Information Technology & People* 7(4): 6–37.
2. Bhanu, Bir, and Hui Chen. 2008. *Human ear recognition by computer*. Springer.
3. Iannarelli, A. 1989. *Ear identification*. Fremont California: Paramount Publishing Company.
4. Chang, Kyong, Kevin W. Bowyer, Sudeep Sarkar, and Barnabas Victor. 2003. Comparison and combination of ear and face images in appearance-based biometrics. *IEEE Transactions on Pattern Analysis and Machine Intelligence* 25(9): 1160–1165.
5. Yan, Ping, and K.W. Bowyer. 2007. Biometric recognition using 3D ear shape. *IEEE Transactions on Pattern Analysis and Machine Intelligence* 29(8): 1297–1308.

Chapter 2
Ear Detection in 2D

2.1 Introduction

Most of the well known ear biometric techniques have focussed on recognition on manually cropped ears and have not used automatic ear detection and segmentation. This is due to the fact that detection of ears from an arbitrary profile face image is a challenging problem as ear images may vary in scale and pose (due to in-plane and out-of-plane rotations) under various viewing conditions. However, for an efficient ear recognition system, it is desired to detect the ear from the profile face image in an automatic manner.

There exist few techniques in the literature which can be used to detect ear automatically. A detailed review of these techniques is as follows. The first well known technique for ear detection is due to Burge and Burger [1]. It has detected ears with the help of deformable contours. But contour initialization in this technique needs user interaction. As a result, ear localization is not fully automatic. Hurley et al. [2] have used force field technique to get the ear location. The technique claims that it does not require exact ear localization for ear recognition. However, it is only applicable when a small background is present in ear image. In [3], Yan and Bowyer have used manual technique based on two-line landmark to detect ear where one line is taken along the border between the ear and the face while other line is considered from the top of the ear to the bottom. The 2D ear localization technique proposed by Alvarez et al. [4] uses ovoid and active contour (snake) [5] models. Ear boundary is estimated by fitting the contour of an ear in the image by combining snake and ovoid models. This technique requires an initial approximated ear contour to execute and hence cannot be used in fully automated ear recognition system. There is no empirical evaluation of the technique.

Yan and Bowyer [6] have proposed another technique by considering a predefined sector from the nose tip as the probable ear region. It first computes the ear pit using the curvature information obtained from 3D data and uses its boundary to initialize active contour which detects the ear boundary. It fails if the ear pit is occluded. It produces 78.79 % correct ear segmentation when only color information is used for

© Springer Science+Business Media Singapore 2015
S. Prakash and P. Gupta, *Ear Biometrics in 2D and 3D*,
Augmented Vision and Reality 10, DOI 10.1007/978-981-287-375-0_2

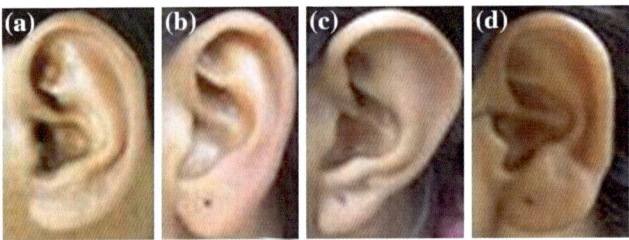

Fig. 2.1 Different ear shapes. **a** Round, **b** oval, **c** triangular, **d** rectangular

active contour conversion. Ansari and Gupta [7] have presented an ear detection technique based on edges of outer ear helices. The accuracy of the this technique is reported to be 93.34 % on 700 sample images collected at IIT Kanpur. The technique solely relies on the parallelism between the outer helix curves and does not use any structural information present in inner part of the ear and hence, it may fail if the helix edges are poor. Yuan and Mu [8] have proposed a technique based on skin-color and contour information. It detects ear by roughly estimating the ear location and by improving the localization using contour information. It considers ear shape elliptical and fits an ellipse to the edges to get the accurate position of the ear. There is no quantitative evaluation reported for the technique.

Another ear localization technique which exploits the elliptical shape of the ear has been proposed in [9]. It has been tested on 252 images of 63 individuals selected from XM2VTS [10] and 942 image pairs of 302 subjects of UND database. For XM2VTS database which is relatively small and has less complex images, the technique has achieved 100 % detection rate. However for UND database which contains complex images, it has offered only 91 % detection rate. Moreover, the assumption of considering ear shape elliptical for all subjects may not be true and hence, may not help in detecting the ear, in general. For example, as shown in Fig. 2.1, assumption of elliptical boundary may correctly approximate the ear boundaries for round and oval shapes but may fail in case of triangular and rectangular shapes. Also, this assumption restricts the ear localization to a controlled environment as the presence of background objects may produce false positives.

In [11], Sana et al. have given a template based ear detection technique where to detect ears at different scales, ear templates of different sizes are maintained. In practice, any predefined set of templates may not be able to handle all situations. Experimental study in this technique has used 1800 images collected from 600 individuals. However ear detection accuracy is not reported explicitly in the paper. In [12, 13], there are two techniques for ear localization which are also based on template matching. In these techniques, an ear template which is created off-line is resized to obtain a template of suitable size. Resizing is done using the size of the skin part of profile face image which works well when profile face includes only facial parts. But while capturing the profile face, an image may include other skin parts such as neck. This makes the size of the skin area larger than the actual and leads to an incorrect resizing of the ear template and hence, it produces an erroneous ear localization. Techniques in [12, 13] have been tested on part of IIT Kanpur ear database containing profile face images of 150 individuals and found to have accuracy of 94 % and 95.2 % respectively.

Attarchi et al. [14] have proposed an ear detection technique based on the edge map. It relies on the hypothesis that the longest path in edge image is the outer boundary of the ear. It works well only when there is small background present around the ear and fails if ear detection is carried out in whole profile face image. Performance of ear detection of this technique has been reported on two databases, namely USTB database which contains 308 ear images from 77 persons [15] and Carreira-Perpinan database which includes 102 ear images from 17 persons [16]. Accuracy has been found to be 98.05 % for USTB database and 97.05 % for Carreira Perpinan database. A cascaded AdaBoost based ear detection approach has been proposed in [17]. The technique uses Haar-like rectangular features as the weak classifiers. AdaBoost is used to select good weak classifiers and then to combine them into strong classifiers. A cascade of classifiers is built which works as the final detector. The detection performance of the cascaded ear detector has been evaluated for 203 profile face images of UND database and is reported to have accuracy of 100 %. However, the technique needs huge amount of time for training and has been tested on relatively small set of images.

In [18], an ear localization technique has been proposed which is based on hierarchical clustering of the edges. To identify the edge cluster related to ear, the technique assumes approximate size of the ear cluster. Because of this, it works well when scale of the profile face image does not vary much. The technique is rotation invariant. However to handle scale, cluster size of the ear needs to be adjusted which may not be possible without user intervention. The technique has been tested on a database consisting of 500 profile face images of human profile faces collected at IIT Kanpur and found to have an accuracy of 94.6 %.

In [19], an ear detection technique using the image ray transform has been presented. The transform is capable of highlighting the tubular structures of the ear such as helix. The technique exploits the elliptical shape of the helix to perform the ear localization. However, assumption of ear shape being elliptical may be very rigid. The technique has achieved 99.6 % ear detection on 252 images of the XM2VTS database [10]. Ibrahim et al. [20] have employed a bank of curved and stretched Gabor wavelets (popularly called banana wavelets) for ear detection. A 100 % detection rate is achieved by this technique on images of XM2VTS database. In [21], a technique for ear detection has been presented by Kumar et al. where skin-segmentation and edge detection has been used for initial rough ear region localization. Region based active contour technique [22] has been further applied to get exact location of ear contours. The technique has been tested on 700 ear images and has achieved 94.29 % correct ear detection. This technique is applicable only when small background is present in the ear images. It can be observed that most of the techniques discussed above which have achieved almost 100 % correct ear detection rate have been tested on small data sets (≤300 images).

Most of these techniques can detect the ear only when a profile face image contains a small background around the ear. These techniques are not very efficient, particularly when profile face images are affected by scaling and rotation (pose variations). Moreover, they are not fully automatic and fast enough to be deployed in realtime applications. However, it is often required, specially in non-intrusive applications, to detect the ear from a whole profile face image which may be affected by scale and pose variations.

This chapter discusses an efficient ear localization technique which attempts to address these issues. The technique is invariant to scale, rotation and shape. It makes use of connected components of a graph constructed with the help of edge map of the profile face image to generate a set of probable ear candidates. True ear is detected by performing ear identification using a rotation, scale and shape invariant ear template.

Rest of the chapter is organized as follows. Section 2.2 briefly describes a skin color used for skin segmentation and Speeded Up Robust Features (SURF) used in ear template creation in the ear detection technique discussed in this chapter. Next section presents the ear detection technique. Rotation, scale and shape invariance of this technique has been discussed in Sect. 2.4. Experimental results are analyzed in Sect. 2.5.

2.2 Preliminaries

2.2.1 Color Based Skin Segmentation

This section presents a color based technique to segment skin and non-skin regions. It is similar to the skin segmentation technique proposed in [23] which has used 1976 CIE Lab color space for image representation. However, we have represented images in YCbCr space because it is perceptually uniform [24] and is widely used in video compression standards such as JPEG and MPEG [25].

The technique is capable of adapting different skin colors and lighting conditions. It performs skin segmentation in *YCbCr* color space as it is more suitable for characterizing skin colors. It first converts an image from *RGB* color space to *YCbCr* color space and then uses *YCbCr* color information for further processing. In *RGB* color space, *(R, G, B)* components represent not only color information but also luminance which may vary across a face due to the ambient lighting. This makes *(R, G, B)* components an unreliable measure for separating skin from non-skin regions. *YCbCr* color space separates luminance from the color information and hence, provides a way to use only color information for segmenting skin and non-skin regions.

The distribution of skin colors of different people is found to be clustered in a small area in the *YCbCr* color space. Although skin colors of different people may vary over a wide range, they differ more in brightness than its color. Due to this fact, skin color model is developed in *YCbCr* color space and only chrominance components (*Cb* and *Cr*) are used for modeling the skin pixels. Since color histogram of skin color distribution of different people is clustered at one place in *Cb, Cr* plane, it can be represented by a Gaussian model $N(\mu, \Sigma)$ with mean μ and covariance Σ.

With the Gaussian fitted skin color model, likelihood of skin for each pixel can be computed. If a pixel, having transformed from *RGB* color space to *YCbCr*, has a chromatic color vector $x = (Cb, Cr)^T$, the likelihood $P(x)$ of skin for this pixel can then be obtained by

$$P(x) = \frac{1}{\sqrt{2\pi|\Sigma|}} \exp[-\frac{1}{2}(x - \mu)\Sigma^{-1}(x - \mu)^T] \tag{2.1}$$

Likelihood values obtained in Eq. 2.1 can be used to segment skin and non-skin regions. An adaptive thresholding process [23] is applied on likelihood image (obtained using skin likelihood values for all pixels) to compute an optimal threshold. Skin segmentation is obtained by thresholding the skin likelihood image using this threshold.

2.2.2 Speeded Up Robust Feature Transform

Speeded Up Robust Features (SURF) [26, 27] is a scale and rotation invariant interest point detector and descriptor. It has been designed for extracting highly distinctive and invariant feature points (also called interest points or key-points) from images. The reason behind using SURF for feature representation in this chapter (and also in Chaps. 3 and 5) is that it provides good distinctive features and at the same time is found to be more robust with respect to change in view point, rotation and scale, illumination changes and occlusion [27] as compared to other scale and rotation invariant shape descriptors such as SIFT [28] and GLOH [29].

There are two important steps involved in extracting SURF features from an image. These are finding key-points and computation of their respective descriptor vectors.

2.2.2.1 Key-Point Detection

SURF identifies salient feature points in the image called key-points. It makes use of hessian matrix for key-point detection. For a given point $P(x, y)$ in an image I, the hessian matrix $H(P, \sigma)$ at scale σ is defined as:

$$H(P, \sigma) = \begin{bmatrix} L_{xx}(P, \sigma) & L_{xy}(P, \sigma) \\ L_{yx}(P, \sigma) & L_{yy}(P, \sigma) \end{bmatrix}$$

where $L_{xx}(P, \sigma)$, $L_{xy}(P, \sigma)$, $L_{yx}(P, \sigma)$ and $L_{yy}(P, \sigma)$ are the convolution of the Gaussian second order derivatives $\frac{\partial^2}{\partial x^2}g(\sigma)$, $\frac{\partial^2}{\partial x \partial y}g(\sigma)$, $\frac{\partial^2}{\partial y \partial x}g(\sigma)$ and $\frac{\partial^2}{\partial y^2}g(\sigma)$ with the image I at point P respectively.

(a)   **(b)**

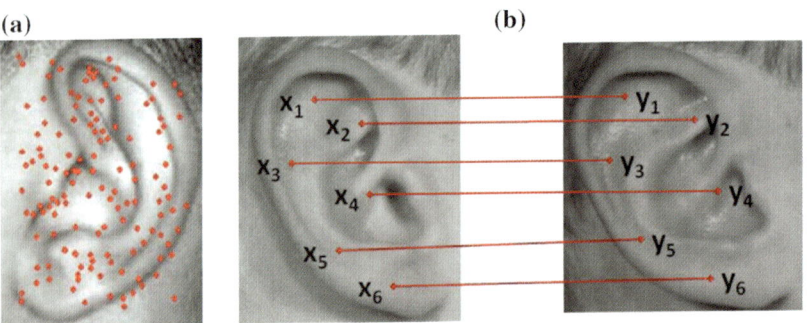

Fig. 2.2 Example of SURF features and matching, **a** SURF feature points, **b** matching

To speed up the computation, second order Gaussian derivatives in Hessian matrix are approximated using box filters. To detect key-points at different scales, scale space representation of the image is obtained by convolving it with the box filters. The scale space is analysed by up-scaling the filter size rather than iteratively reducing the image size. In order to localize interest points in the image and over scales, non-maximum suppression in a $3 \times 3 \times 3$ neighborhood is implemented. Figure 2.2a shows an example where SURF feature points.

2.2.2.2 Key-Point Descriptor Vector

In order to generate the descriptor vector of a key-point, a circular region is considered around each detected key-points and Haar wavelet responses dx and dy in horizontal and vertical directions are computed. These responses are used to obtain the dominant orientation in the circular region. Feature vectors are measured relative to the dominant orientation resulting the generated vectors invariant to image rotation. Also a square region around each key-point is considered and it is aligned along the dominant orientation. The square region is divided into 4×4 sub-regions and Haar wavelet responses are computed for each sub-region. Sum of the wavelet responses in horizontal and vertical directions for each sub-region are used as features. In addition, the absolute values of responses are summed to obtain the information about the polarity of the image intensity changes. Thus, the feature vector V_i for ith sub-region is given by

$$V_i = \{\Sigma dx, \Sigma dy, \Sigma |dx|, \Sigma |dy|\}$$

SURF descriptor vector of a key-point is obtained by concatenating feature vectors V_is from all sixteen sub-regions around the key-point resulting a descriptor vector of length $16 \times 4 = 64$. This is called SURF-64. Extended version of SURF (known as SURF-128) which is more distinctive, adds a couple of more distinctive features to the descriptor vector. It uses the sums same as described above, however splits these values up further. It computes the sum of d_x and of $|d_x|$ separately for $d_y < 0$ and $d_y \geq 0$. Similarly, the sum of d_y and of $|d_y|$ are found according to the sign of d_x, hence doubling the number of features elements in the descriptor vector.

2.2.2.3 Matching in SURF

Matching in SURF is performed using nearest neighbor ratio matching. The best candidate match for a key-point of an image in another image is found by identifying its nearest neighbor in the key-points from the second image where nearest neighbor is defined as the key-point with minimum Euclidean distance from the given key-point of first image with respect to their descriptor vectors. The probability that a match is correct is determined by computing the ratio of distance from the closest neighbor to the distance of the second closest one. A match is declared successful if the distance ratio is less than of equal to a predetermined threshold $\tau \in (0, 1)$. Algorithm for SURF matching is described in Algorithm 2.1. Figure 2.2b shows an example where SURF matching points between two ear images are shown.

Algorithm 2.1 SURF Matching

- **Input:** Two sets of descriptor vectors $D^1 = \{D_1^1, D_2^1, \ldots, D_n^1\}$ and $D^2 = \{D_1^2, D_2^2, \ldots, D_m^2\}$ corresponding to n and m key-points of images I_1 and I_2 to be matched and matching threshold $\tau \in (0, 1)$.
- **Output:** Matching score N stating number of matching descriptor vectors in two images.

1: $N \leftarrow 0$
2: **for** $i = 1$ to n **do**
3:    **for** $j = 1$ to m **do**
4:       Compute $distance[j] = Euclidian\_Distance(D_i^1, D_j^2)$.
5:    **end for**
6:    Compute $[Sort\_Dist, Original\_Index] = Sort(distance)$ where $Sort(.)$ is a function which sorts $distance$ array in ascending order and returns sorted distance values in array $Sort\_Dist$ and their corresponding original index values of $distance$ array in array $Original\_Index$.
7:    **if** $\dfrac{Sort\_Dist[1]}{Sort\_Dist[2]} \leq \tau$ **then**
8:       Descriptor D_i^1 of image I_1 matches to descriptor $D_{Original\_Index[1]}^2$ of image I_2 where $Original\_Index[1]$ is the index of the matched descriptor from image I_2.
9:       $N \leftarrow N + 1$
10:   **end if**
11: **end for**
12: Return matching score N.

2.3 Ear Detection Technique

This section presents an efficient ear detection technique. This technique is based on the fact that in a profile face image, ear is the only part which contains large variation in the pixel intensities, resulting this part rich in edges. This can be visualized from the image shown in Fig. 2.3f which displays the edge image of the skin segmented image of Fig. 2.3e. It can be observed that the ear part has larger edge density as compared to other parts. Further, it can also be noticed that all edges belonging to the ear part contain some curvature. These characteristics are exploited for ear localization in the presented technique which computes edge clusters in the edge map obtained from the profile face image and examines them for ear localization. Flow diagram of the technique is presented in Fig. 2.4.

2.3.1 Preprocessing

Profile face image undergoes a preprocessing phase before ear localization. This involves skin segmentation where skin areas of the image are segmented. Further, edge computation is carried out on skin segmented image. In the next step, obtained

Fig. 2.3 Skin segmentation in profile face image. **a** Input color image, **b** skin-likelihood image, **c** binary image, **d** dilated binary image, **e** skin segmented image, **f** edge image

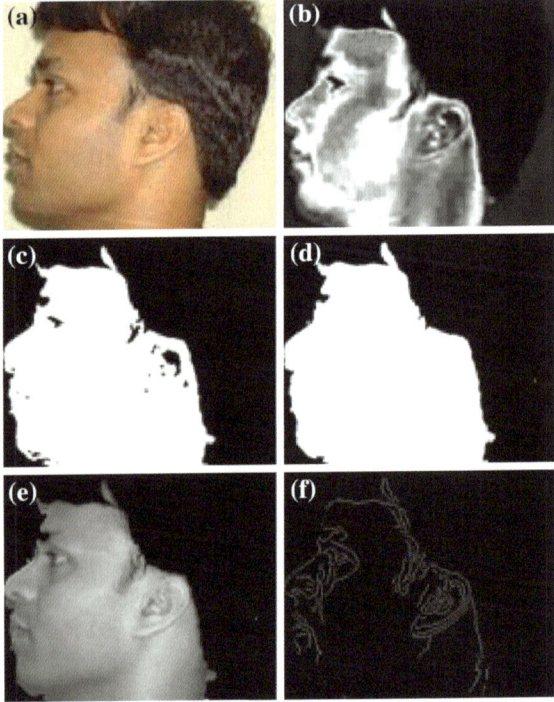

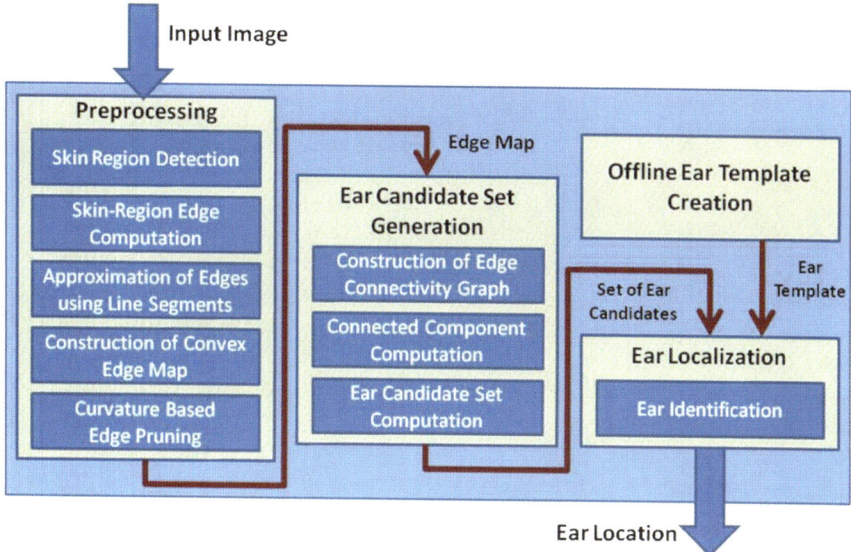

Fig. 2.4 Flow chart of 2D ear detection technique

edges are approximated using line segments and subsequently used in the construction of convex edge map. Erroneous edges are pruned out in the last step.

2.3.1.1 Skin Region Detection

Since ear exist in skin region, non-skin regions of the profile face should be segmented and removed from further processing. The skin color model discussed in Sect. 2.2.1 is used for skin segmentation. It transforms a color image into a gray scale image (called skin-likelihood image) using Eq. 2.1 such that the gray value at each pixel shows the likelihood of the pixel belonging to the skin. With an appropriate thresholding, the gray scale image is further transformed to a binary image segmenting skin (white pixels) and non-skin (black-pixels) regions. Since people with different skins have different likelihood, an adaptive thresholding process [23] is used to achieve the optimal threshold for each image.

The binary image showing skin and non-skin regions may contain some holes in it due to the presence of noise in the profile face image. Dilation is applied to fill these holes before using it for skin segmentation. The effect of this operation is to enlarge gradually the boundaries of regions of foreground pixels (i.e. white pixels). Thus the area of foreground pixels grows while filling holes within regions.

Figure 2.3 considers an example of skin region detection with various intermediate steps. For a color image given in Fig. 2.3a, corresponding skin-likelihood image is shown in Fig. 2.3b. It can be noticed that skin regions in Fig. 2.3b are brighter than the

Fig. 2.5 Edge
approximation: **a** original
edge image, **b** edges
approximated by line
segments

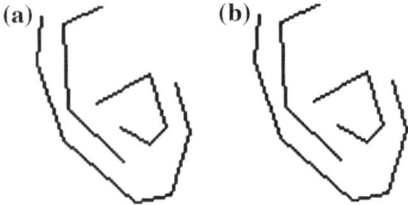

non-skin regions. Figure 2.3c shows the binary image obtained by thresholding the skin-likelihood image. Dilation is applied on this image to repair it by filling small holes present in it. Figure 2.3d shows the repaired binary image. It is used for skin region detection where pixels of the profile face image corresponding to white pixels of the binary image are considered as skin pixels. Figure 2.3e shows the final skin segmented image. It can be observed from segmentation result that not all detected skin regions contain ear. Hence, ear localization can be used to locate the ear in all these skin like segments.

2.3.1.2 Skin-Region Edge Computation

Edge detection is carried out on skin segmented image using Canny edge operator and a list of all edges is computed. An edge in the list is obtained by connecting edge points together into a sequence of pixel coordinate pairs. Wherever an edge junction[1] is encountered, the sequence is terminated and a separate edge point sequence is generated for each of the branches and added to the list. This generates a set of edges containing two end points. Let χ be the set of all such edges.

2.3.1.3 Approximation of Edges Using Line Segments

All pixels present in an edge (belonging to set χ) may not be equally important and may not be necessarily required to represent the edge. So to remove redundant pixels from an edge and to get its compact representation, an edge is approximated using a set of line segments which keeps only those pixels which are important.

Line segments for an edge (belonging to set χ) can be found by considering array of edge points and finding the size and position of the maximum deviation from the line that joins the endpoints of the edge. If the maximum deviation exceeds the allowable tolerance, the edge is shortened to the point of maximum deviation and the process is repeated. In this manner each edge is broken into line segments, each of which adheres to the original data with a specified tolerance. Figure 2.5b shows an example of edge approximation by line segments for the edge image in Fig. 2.5a. Let χ_{ls} be the set containing all edges obtained after line segments fitting.

---

[1] Edge junction is a pixel where an edge divides into two or more edges.

2.3.1.4 Construction of Convex Edge Map

It is observed that edges belonging to the ear have convex[2] nature. However, because of the presence of noise such as hair near the ear, often false edges join true ear edges and make them non-convex. It may lead to an improper ear localization. This usually happens with the outer helix edges of the ear. To avoid this, the derived edges with set χ_{ls} are broken into a set of convex edges. Let χ_{convex} be the set of all convex edges. Identification of convex and non-convex edges and breaking the non-convex edges into convex can be done as follows.

Algorithm 2.2 Construction of Convex Edge Map

- **Input:** Set χ_{ls} of edges approximated with line segments.
- **Output:** Set χ_{convex} of convex edges.

1: Define a null set χ_{convex}.
2: **for** $\forall e \in \chi_{ls}$ **do**
3:    Compute ρ_e using Eq. 2.2.
4:    **if** $\rho_e == 0$ **then**
5:       Add e to χ_{convex}.
6:    **else**
7:       Break e into a set of convex edges and add these edges to χ_{convex}.
8:    **end if**
9: **end for**

Let there be an edge $e \in \chi_{ls}$ obtained after approximation. Assume e consists of k line segments with ith line segment, l_i, having end points: t_i and t_{i+1}. Let the line segment l_i be represented by vector $\overrightarrow{v}_i = t_{i+1} - t_i$. Let $\overrightarrow{v}_{i,i+1}$ be the vector cross-product of $\overrightarrow{v}_i$ and $\overrightarrow{v}_{i+1}$ (vector representing line segment l_{i+1}). The edge e is convex if directions of $\overrightarrow{v}_{i,i+1}$, for all i, are found to be same. To test whether an edge e is convex or non-convex, a decision parameter ρ_e can be estimated as follows.

$$\rho_e = \begin{cases} 0, \text{ if directions of vectors } \overrightarrow{v}_{(i,i+1)}, \forall i, \text{ are same} \\ 1, \text{ otherwise} \end{cases} \quad (2.2)$$

The edge e is convex if ρ_e is 0. To break a non-convex edge into a set of convex edges, it is scanned from one end to another and direction of each cross-product is analyzed. When a cross-product is found to be of different direction with respect to the previous cross-product, the edge is broken at that point. This procedure is continued till whole edge is broken into convex edges. Steps for construction of convex edge map are given in Algorithm 2.2.

Figure 2.6 presents an example of breaking of edges into convex type. Figure 2.6a shows two edges, one convex (edge $ABCD$) and another non-convex (edge $PQRS$). Vector representation of the line segments used in these edges and the direction of

---

[2] Edges which have curvature throughout either positive or negative are considered convex.

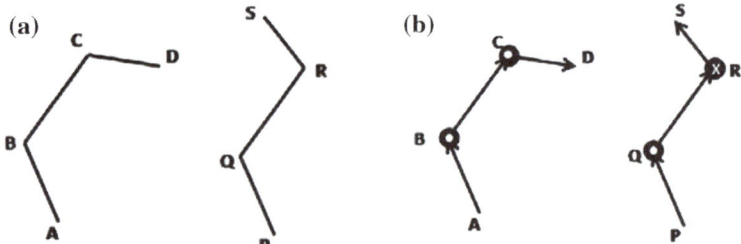

Fig. 2.6 An example of breaking edges into convex type: **a** edge map, **b** line segments with vector representation

the cross-products for adjacent vectors have been shown in Fig. 2.6b. Circle with a cross and circle with a dot at the joining points of two vectors represent the outward and the inward directions of the cross-product respectively. In edge *ABCD* of Figure 2.6b, it can be observed that all cross-products are inward so this edge is marked as convex while in edge *PQRS* of Fig. 2.6b, one cross-product is inward and other is outward so the edge is marked as non-convex. While scanning the edge *PQRS* from lower end, direction of the cross-product at point *R* is found to be different from the previous direction of the cross-product, so the edge *PQRS* is broken at point *R* into two edges: *PQR* and *RS*.

Ear localization accuracy can be improved by converting all non-convex edges to convex type. Breaking of non-convex edges into convex helps in removing the outlier edges (created due to noise). If the edges are converted to convex type, while constructing the edge connectivity graph, most of the outlier edges get isolated and do not appear in the connected component representing the ear and hence, do not affect the ear localization result. Figure 2.7 shows one such example of ear detection. In Fig. 2.7a, edge marked as *A* contains some erroneous part at its lower end arose due to the linking of true ear edge to a noisy edge present in the neck part. Due to this, when the edge *A* participates in the connected component representing ear, localization result includes some skin portion from the neck which does not belong to the ear. Figure 2.7c shows the localization result for this. When the edge *A* is segmented into convex edges *B* and *C* (Fig. 2.7b), lower part of the edge *A* (i.e. *B* after breaking) gets isolated from the ear edge cluster and remaining ear edge cluster produces the correct localization result. Figure 2.7d shows the localization result for this.

Any noise mainly affects the outer edge (helix) of the ear and hence, conversion of non-convex edges to convex primarily helps to remove noisy edges from the outer helix. Since detection of outer helix edge is difficult and computationally expensive, in the technique all edges are converted to convex type. However, conversion of non-convex edges present in the inner parts of the ear to convex type does not have any impact on the localization performance.

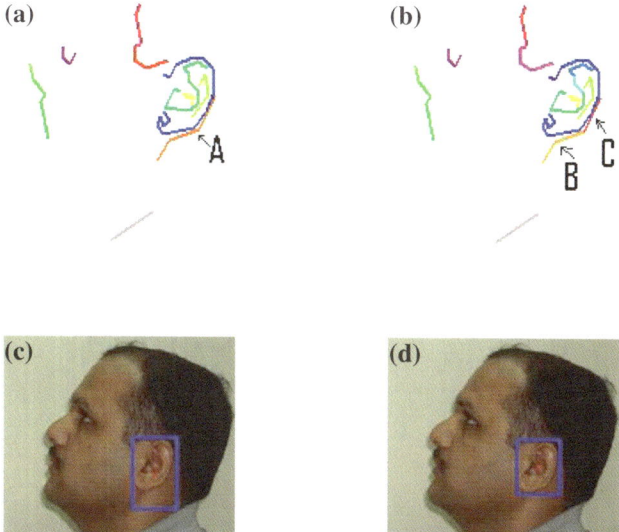

Fig. 2.7 Ear detection by breaking non-convex edges into convex edges where colors in (**a**) and (**b**) used to differentiate edges. **a** Edges Before convex type segmentation, **b** edges after convex type segmentation, **c** detected ear when (**a**) is used, **d** detected ear when (**b**) is used

2.3.1.5 Curvature Based Edge Pruning

All edges in the set χ_{convex} are of convex nature and are represented by line segments. It can be seen that each edge in the set χ_{convex} represented by one line segment (or two points) depicts a linear edge in the original edge map (set χ). Since all edges belonging to the ear contain some curvature, they need more than one line segment (or more than two points) for their representation. In other words, all edges having two points cannot be the part of ear edges and hence can be removed from the set χ_{convex}. This results a new edge set χ_c containing only the edges belonging to ear. Set χ_c can be formally defined as: $\chi_c = \{e \mid e \in \chi_{convex} \text{ and } \gamma(e) > 2\}$, where $\gamma(e)$ gives the number of points used in edge e to approximate it by line segments.

2.3.2 Ear Candidate Set Generation

This phase builds an edge connectivity graph which is used to find the connected components in the graph to obtain ear candidate set.

2.3.2.1 Building Edge Connectivity Graph

The set χ_c can be used to define the edge map of the profile face image. Let there be n edges in χ_c. The ith edge e_i in χ_c is defined by a point p_i. Thus χ_c can be represented by a set P of points $p_1, p_2, \ldots, p_n$ where p_i refers to e_i for all i. Against each edge e_i, a convex hull[3] $CH(e_i)$ is defined. If two convex hulls $CH(e_i)$ and $CH(e_j)$ intersect each other, then points p_i and p_j are connected through an arc[4] of a newly defined graph $G = (V, E)$ with the set of vertices V and the set of edges E, where

$$V = \{p_i \mid p_i \in P\}$$
$$E = \{(p_i, p_j) \mid CH(e_i) \text{ intersects } CH(e_j)\}$$

G is called edge connectivity graph. Algorithm 2.3 provides the steps invoked in building the graph G.

Algorithm 2.3 Construction of Edge Connectivity Graph

- **Input:** Edge map χ_c of profile face image I.
- **Output:** Edge connectivity graph $G = (V, E)$.

1: Define a graph $G = (V, E)$ where V and E are initially null.
2: Define a set $P = \{p_1, p_2, \ldots, p_n\}$ for the n edges in set χ_c such that point p_i represents ith edge e_i in set χ_c.
3: Define $V = \{p_i | p_i \in P\}$.
4: Define convex hull CH_i for each edge e_i, $e_i \in \chi_c$.
5: **for all** $i, j \in [1, n]$ **do**
6:    **if** $CH(e_i)$ intersects $CH(e_j)$ **then**
7:       Connect points p_i and p_j by an edge (p_i, p_j) in graph G and add it to E.
8:    **end if**
9: **end for**
10: Return G.

One can observe that the ear edges are mostly convex in nature and if one moves from outer part of the ear towards inside, then most of the outer edges contain inner ear edges. Due to this nature of ear edges, convex hulls of the outer edges intersect the convex hulls of the inner edges. This almost guarantees that the convex hull of an edge belonging to the ear intersects at least another convex hull of the edge belonging to the ear. So this criterion to define connection between vertices (points) in a graph connects (directly or indirectly) all vertices belonging to the ear part with each other. Moreover, this criterion can define the connectivity irrespective of the scale; as a result, it makes the technique scale invariant. In general, property of one

---

[3] Convex hull for an edge is a tightest convex polygon which includes all edge points.

[4] In this work, "arc" signifies the notion of an edge in a graph. The word "edge" is used in the context of an edge in an image which is a set of connected pixels representing points of high intensity gradient in the image.

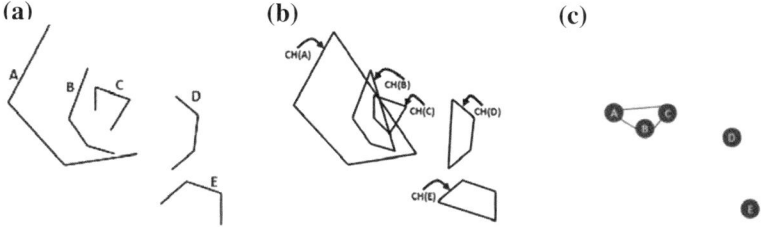

Fig. 2.8 An example of construction of edge connectivity graph. **a** Synthetic edges, **b** convex hulls, **c** connectivity graph

edge containing another is not true for the edges belonging to other parts of the profile face image; so vertices corresponding to these edges remain mostly isolated in the edge connectivity graph.

Figure 2.8 shows an example of an edge map and convex hulls of edges. It is seen from Fig. 2.8b that convex hulls of edges A, B and C intersect with each other. So vertices corresponding to these edges are connected to each other in the graph as shown in Fig. 2.8c. Points D and E are isolated in Fig. 2.8c since their respective convex hulls in Fig. 2.8b do not intersect to convex hull of any other edge.

It can be noted that there can be some simple criteria to define the connectivity among the vertices in edge connectivity graph. One such criterion may be based on some distance metrics between two edges. However, such choice makes ear detection scale dependent. This is due to the fact that the distance threshold required to define the connectivity among the vertices may vary for the images of different scales.

2.3.2.2 Connected Component Computation

Two vertices are in the same connected component of an undirected graph if there exists a path between them. After defining the graph for the edge map of profile face image, its connected components are computed. These components are analyzed one by one to localize the ear. To compute the connected components in the graph $G = (V, E)$, we have used a breath first search based algorithm described in [30].

Figure 2.9 presents an example of edge connectivity graph and connected components labeling. Figure 2.9a shows an edge image obtained from a profile face image. A graph, shown in Fig. 2.9b, is constructed for this edge image and connected components (enclosed inside rectangular boundaries) are computed. Magnified view of the component A present in Fig. 2.9b can be seen in Fig. 2.9c.

2.3.2.3 Ear Candidate Set Computation

Ideally, it is believed that the vertices representing ear edges are connected to each other (directly or indirectly) and form one connected component while all other vertices representing non-ear edges remain isolated. Hence the criterion based on

Fig. 2.9 An example of
edge connectivity graph and
connected components
labeling. **a** Edge map (colors
used to differentiate edges),
b graph for the edge map of
(**a**) with connected
components labeling,
c magnified view of
component A of (**b**)

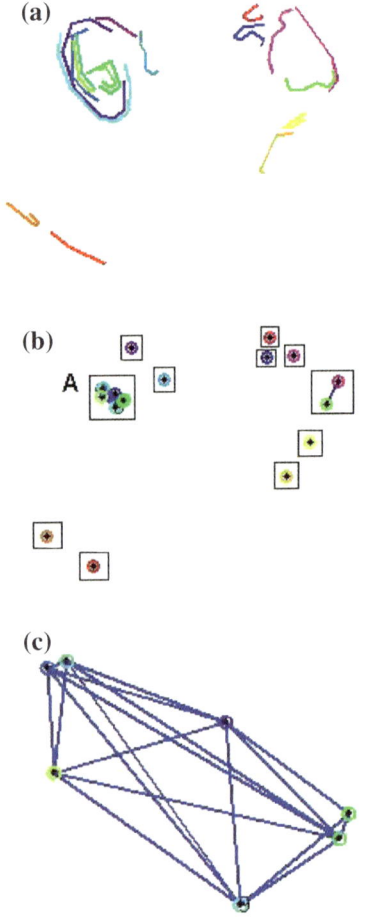

the size of the component can be used to find out the connected component representing ear. However, there may exist few more places in the profile face where due
to noise a convex hull of one edge may intersect that of other edges and give rise to a
large connected component. Hence, each connected component in the edge connectivity graph which has two or more vertices is considered as a probable candidate to
represent the ear. Any connected component having single vertex can be straightaway
removed from the graph as it cannot represent the ear. Let $K = \{K_1, K_2, \ldots, K_m\}$
be the set of connected components of graph G where each component has two or
more number of vertices. Average vertex degree of a connected component K_j is
defined as:

$$d(K_j) = \frac{\Sigma_{i=1}^{n_j} d(p_i)}{n_j} \tag{2.3}$$

where $d(p_i)$ is the degree of vertex p_i and n_j is the total number of vertices present in component K_j. As stated earlier, ear part of the profile face image is rich in edges due to large intensity variation in this region; hence, it is less probable that a connected component representing an ear will have only two vertices (or average vertex degree one). Therefore, to further prune out the false connected components, only the components having average vertex degree greater than one can be considered to obtain probable ear candidates. A probable ear candidate in a profile face image is defined as the image portion which is cropped using the bounding box of the edges participating in a connected component. A set of ear candidates is computed using all connected components satisfying the criterion on the average vertex degree. Algorithm 2.4 presents steps to generate ear candidate set using connected components.

Algorithm 2.4 Computation of Ear Candidate Set

- **Input:** Set $K = \{K_1, K_2, \ldots, K_m\}$ containing m connected components of G.
- **Output:** Set I_E containing the image portions cropped from the profile face which are the probable candidates for ear.

1: **for** $j = 1$ to m **do**
2:     $d(K_j) = \frac{1}{n_j}\Sigma_{i=1}^{n_j}d(p_i)$, $p_i \in K_j$ and $d(p_i)$ is the degree of vertex p_i, n_j is the number of vertices in K_j.
3: **end for**
4: Define set $Q = \{j | d(K_j) > 1\}$
5: Define set $H = \{H_j | j \in Q\}$ where $H_j = \{e_i | p_i \in K_j, j \in Q\}$ contains edges with the edge e_i represented by point p_i in G as discussed in Sect. 2.3.2.1.
6: Define $B = \{B_j | B_j$ is the bounding box of the edges present in $H_j \in H\}$.
7: Obtain $I_E = \{I_j | I_j$ is cropped image from profile face using $B_j \in B\}$.
8: Return probable ear candidate set I_E.

2.3.3 Ear Localization

It is carried out by identifying the true ear among the probable ear candidates with the help of an ear template which is created off-line. The template works as an ear representative which depicts the characteristics of ears of various scales, rotations and shapes. Identification is performed by comparing the probable ear candidates with the ear template.

2.3.3.1 Ear Template Creation

To identify true ear, the template used for ear identification should exhibit the characteristics of scale and rotation invariance. To compute such a template in the presented technique, a shape descriptor which is invariant to rotation and scale, is used. Among several scale and rotation invariant shape descriptors, SURF [27] provides good distinctive features and at the same time it is robust to changes in viewing condition, rotation and scale. Hence it has been used for ear template creation in this technique. As described in Sect. 2.2.2, SURF represents an image by first identifying some unique feature points in it and then by describing them with the help of a feature descriptor vector. For the description of the feature points, SURF uses intensity content within the neighborhood of feature point and describes it by using the sum of approximated 2D Haar wavelet components.

The ear template is computed by fusing the SURF feature descriptors obtained from various ear images together considering the redundant features only once. Let n be the number of ear images used for template creation. Let $T_1, T_2, \ldots, T_n$ be the SURF feature descriptor sets obtained from these images. A fused ear template T is obtained by

$$T = \bigcup_{i=1}^{n} T_i \qquad (2.4)$$

If the set T_i contains c_i feature descriptor vectors, then total number of descriptor vectors c in T satisfies the following inequality

$$c \leq \sum_{i=1}^{n} c_i \qquad (2.5)$$

Fusion of the SURF feature descriptor sets proceeds incrementally where first two sets T_1 and T_2 are fused to generate a new intermediate feature descriptor set which is further fused with feature descriptor set T_3. This process is continued till all sets are fused together. While fusing two SURF feature descriptor sets T_i and T_{i+1}, SURF matching (described in Algorithm 2.1) is performed between the two sets to find out the redundant feature descriptor vectors. If a descriptor vector in a set matches to a descriptor vector in another set, it is considered as common to both and is used only once in fusion. For example take the reference of Fig. 2.2b, if a feature point x_i from the first ear image matches to a feature point y_i in the second ear image, either descriptor vector for x_i or descriptor vector for y_i is used in fused feature descriptor set.

A SURF descriptor vector can be either of 64 dimensions or of 128 dimensions. A 128 dimensional descriptor vector provides more discriminative power as compared to 64 dimensional vector, however it involves more computation time. Since an ear template is used to discriminate between ear and non-ear candidates, experimentally it is found that it is sufficient to use 64 dimensional descriptor vector to create a good ear template.

It can be noted that attempts have been made to utilize the power of invariant feature points in other ear biometric systems as well. For example, Bustard and Nixon [31] have used Scale Invariant Feature Transform (SIFT) [28] feature points for registration of probe and gallery image before matching to perform ear recognition.

2.3.3.2 Ear Identification

Let the ear candidate set be $I_E = \{I_1, I_2, \ldots, I_\eta\}$ where η is the cardinality of set I_E and I_k is the image portion of the profile face image representing kth probable ear candidate, $k = 1, 2, \ldots, \eta$. For identification purpose, SURF feature descriptor set is computed for all the ear candidates in I_E. Identification of true ear is performed by comparing the ear template with the SURF descriptor sets of the ear candidates in I_E. Comparison between two SURF descriptor sets is performed using SURF matching which uses the ratio-matching scheme [28] to find out the number of descriptor vectors matching between the two sets. Let D_i and D_j be two descriptor vectors from sets S_1 and S_2 respectively. Let $d(D_i, D_j)$ be a distance metric between the descriptor vectors D_i and D_j. The descriptor vector D_i is said to be matched with D_j if

$$d(D_i, D_j) < \rho \times d(D_i, D_k), \ D_k \in S_2, k \neq j, \forall k \qquad (2.6)$$

where ρ is a constant lying between 0 and 1. A small value of ρ gives a tighter matching while a large value of ρ provides a relaxed matching.

Let $T_E = \{T_{I_1}, T_{I_2}, \ldots, T_{I_\eta}\}$ be the SURF feature descriptor sets for the ear candidate images in I_E. To obtain the true ear, SURF matching is performed between ear template (T) and all elements of T_E and a match score vector *MatchScore* is generated. SURF matching between two descriptor sets returns the number of matched points between them. The true ear candidate I_ξ is obtained such that

$$\xi = \arg \max_i \{MatchScore[i]\}$$

That means, the ear candidate from I_E for which SURF match score is maximum, is declared as the true ear candidate. Algorithm 2.5 provides steps involved in ear identification process.

Algorithm 2.5 Ear Identification using SURF Descriptive Ear Template

---

- **Input:** Set $I_E = \{I_1, I_2, \ldots, I_\eta\}$ containing η probable ear candidates and off-line created ear template T.
- **Output:** I_ξ which is the true ear.

1: Define set $T_E = \{T_{I_1}, T_{I_2}, \ldots T_{I_\eta}\}$ where T_{I_i} represents SURF feature descriptor set for ear candidate $I_i \in I_E$.
2: **for** $i = 1$ to η **do**
3:    $MatchScore[i] = $ SURFmatch(T, T_{I_i}).
4: **end for**
5: $\xi = \arg \max_i \{MatchScore[i]\}$.
6: Return I_ξ.

2.4 Scale, Rotation and Shape Invariance

In the technique discussed here, there are two major steps which play key role in ear localization. First step is the construction of edge connectivity graph which is used to detect probable ear candidates while second one is the identification of true ear among probable ear candidates using ear template. Construction of edge connectivity graph is made scale invariant by defining the connectivity among the vertices in the graph using intersection of convex hulls of corresponding edges. Such criterion to define the connectivity is unaffected by scale changes. Also, intersection of two convex hulls is unaffected if both are rotated; hence rotation also does not influence the process of defining the connectivity of two vertices in the graph. It can be observed that there is no significance of shape invariance at this step.

Rotation, scale and shape invariance at ear identification step is obtained by defining an ear template which exhibits these properties. It is achieved by using SURF feature descriptor for ear template creation which provides rotation and scale invariant description of ear feature points. An ear template is defined as a collection of rotation and scale invariant descriptor vectors obtained from multiple training ear images. Shape invariance is achieved by choosing the ears of different shapes from the database to define the ear template.

2.5 Experimental Results

The technique discussed here has been tested on three databases, namely IIT Kanpur (IITK) database and University of Notre Dame database—Collections E and J2 [32].

2.5.1 Estimation of Parameters

Parameters used for skin segmentation are computed for each data set separately by collecting few skin samples from each of them. Table 2.1 summarizes these parameters for various data sets. Minimum and maximum thresholds used in Canny edge detector are 0.0 and 0.1 respectively while standard deviation of the Gaussian filter σ is set to 1.0 for IITK database and 2.0 for UND database. Value of σ is kept little high for UND database as images in it are noisy. Distance tolerance for edge approximation is set to 20 for both the databases.

Ear template for each data set of IITK and UND databases has been created separately as the nature of the data present in each of them is entirely different. Few images are randomly selected from each data set to compute ear templates. It is found that 50 images from a data set are sufficient to capture the properties of the ears for creating a good ear template. The ratio value ρ used in SURF matching for template creation is taken as 0.5 whereas for true ear identification, it is set to 0.7. Since for

Table 2.1 Gaussian parameters used for skin segmentation in IITK and UND databases

Data set	Mean (Cb, Cr)	Covariance (Cb, Cr)
IITK data sets 1, 2 and 3	$\begin{pmatrix} 102.35 \\ 154.52 \end{pmatrix}$	$\begin{pmatrix} 71.76 & 9.95 \\ 9.95 & 111.77 \end{pmatrix}$
UND-E data set	$\begin{pmatrix} 90.65 \\ 170.23 \end{pmatrix}$	$\begin{pmatrix} 55.55 & -4.79 \\ -4.79 & 107.19 \end{pmatrix}$
UND-J2 data set	$\begin{pmatrix} 109.48 \\ 148.31 \end{pmatrix}$	$\begin{pmatrix} 55.74 & 41.76 \\ 41.76 & 93.62 \end{pmatrix}$

template creation, SURF matching is performed between the ear images, a lower value of ρ (which gives tighter matching) helps in capturing the distinct features of the ears. Ear identification is used to discriminate ear and non-ear candidates and hence matching is relaxed and little higher value of ρ is used.

2.5.2 Results

Figure 2.10 provides the results obtained at various steps of the ear detection for three profile face images taken from UND database. It shows the original input images, profile face edge maps approximated with lines, edge connectivity graph and ear detection results.

Figure 2.11a shows the ear detection results for Data Set 1 of IITK database which contains normal frontal ear images. To show the rotation (pose) and scale invariance of the technique, Data Set 2 of IITK database is used. Figure 2.11b gives few results from Data Set 2 where ears of different sizes and rotations are efficiently detected without any user intervention and change of parameters. The technique has also detected ears successfully for the images of Data Set 3 of IITK database (where images contain out-of-plane rotations) even for the extreme poses ($-40°$ and $+40°$). Figure 2.12a shows detection results for few images taken from IITK Data Set 3. Further, few ear localization results for extreme poses ($-40°$) where ear localization is found to be very challenging are shown in Fig. 2.12b. The technique has localized ears precisely for almost all extreme cases. It has also detected ears of all shapes (viz. round, oval, triangular, rectangular) successfully.

Table 2.2 summarizes ear detection results for IITK database. It is seen that accuracy for Data Set 1 is the highest as it contains frontal ear images. In such images, full ear structure is visible and good amount of edges are obtained which help in achieving strong connectivity among the edges representing ear. Accuracy for Data Set 2 is comparable with that of Data Set 1, in spite of images having variations in scale and rotation. This is due to the fact that the presented technique exploits the structural details of the ear which do not change with scale and rotation. Data Set 3

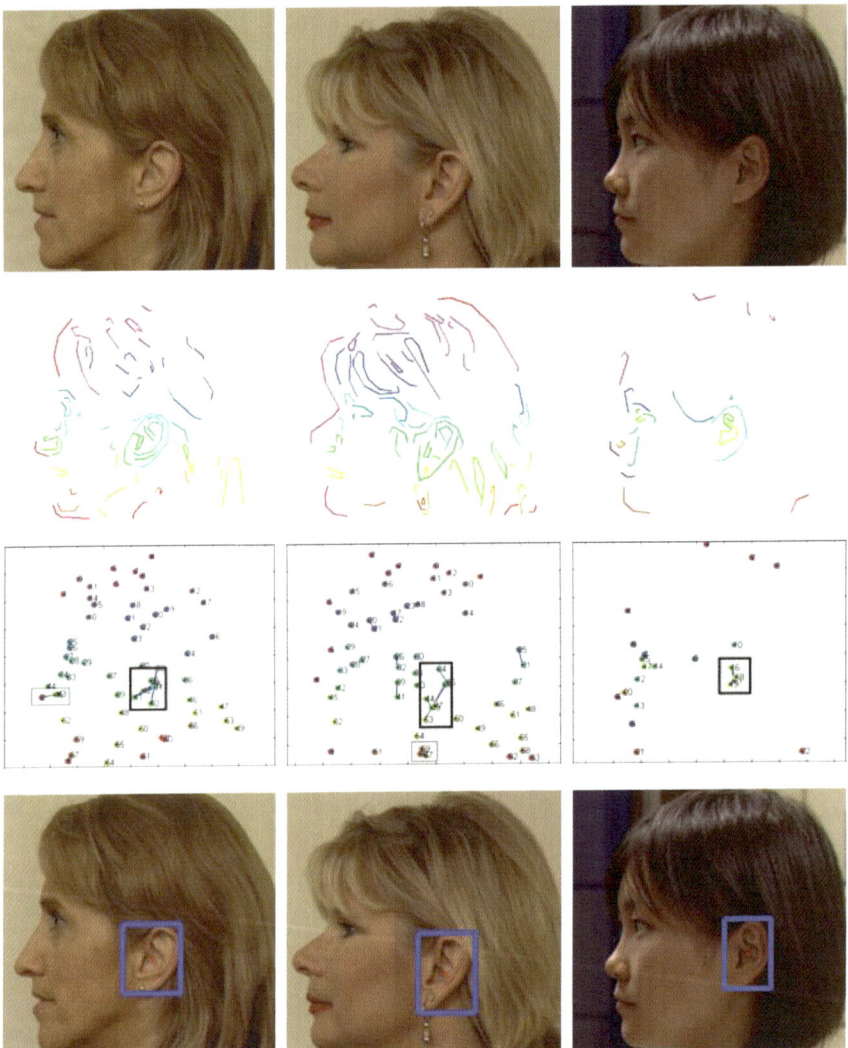

Fig. 2.10 Ear detection: (row-1) original input images, (row-2) edge maps approximated with lines (colors used to distinguish edges), (row-3) edge connectivity graphs (graph components having average vertex degree >1 enclosed in rectangles), (row-4) ear detection results

shows the least accuracy among all data sets of IITK database. This is because in the presence of out-of-plane rotation, the availability of the structural details of the ear decreases as camera moves away from the frontal position. Ear localization results for IITK database are also compared in Table 2.2 with the results reported in [33]. It is evident that the presented technique performs much better than the technique discussed in [33]. This improvement is achieved due to following reasons.

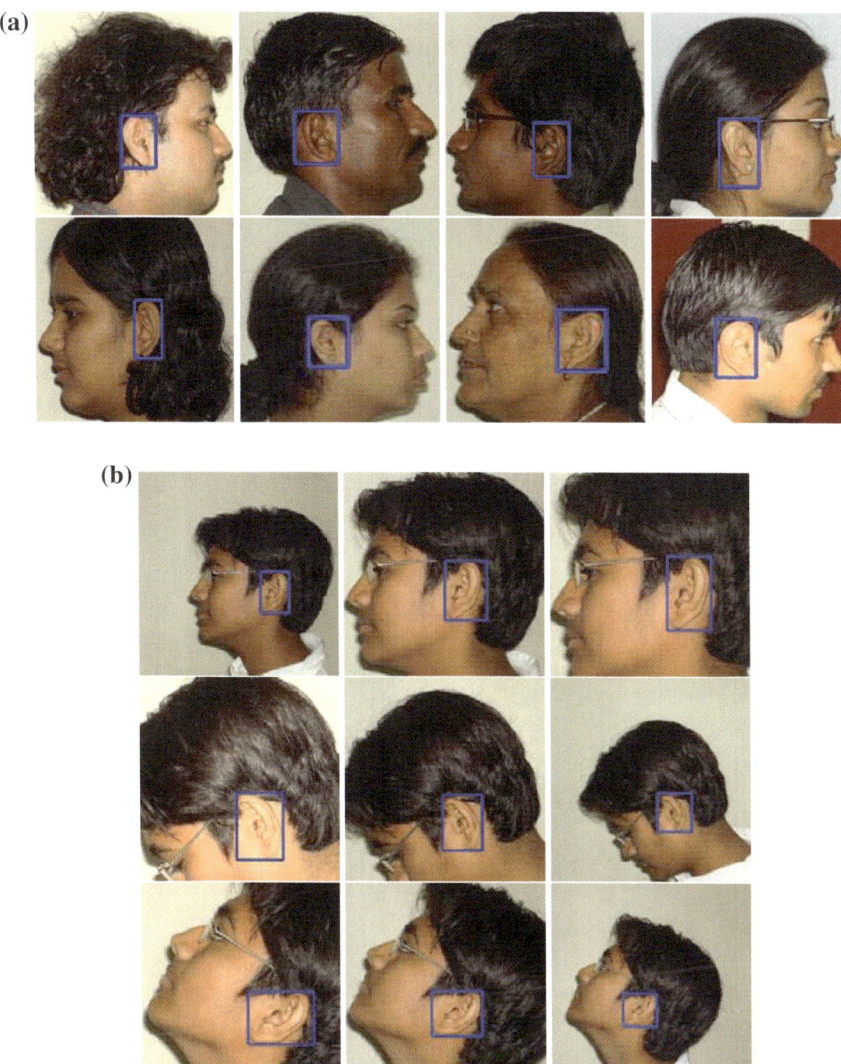

Fig. 2.11 Ear detection results for IITK database. **a** Data set 1, **b** data set 2

1. This technique breaks the derived edges of the profile face into a set of convex edges to reduce the participation of noisy edges in the cluster of true ear edges.
2. This technique has used a rotation, scale and shape invariant ear template which depicts the characteristics of ears of various scales, rotations and shapes. Identification of the true ear is performed by comparing the probable ear candidates with the ear template. Use of rotation, scale and shape invariant ear template greatly helps in localization of ears of various poses, scales and shapes.

(a)

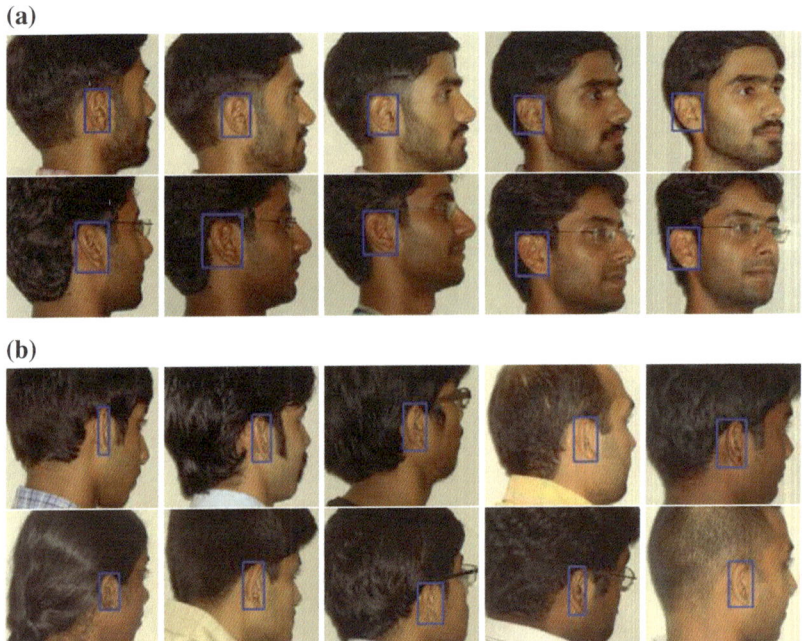

(b)

Fig. 2.12 Ear detection results for IITK database (data set 3). **a** Detection results for two subjects, **b** detection in extreme Views

Table 2.2 Percentage accuracy for IITK database

Data set	# of test images	Ear localization accuracy (%)	
		Reported in [33]	Method discussed here
Data set 1	801	95.88	**99.25**
Data set 2	801	94.73	**98.50**
Data set 3	1070	91.11	**95.61**

3. Identification of true ear among the probable ear candidates with the help of an ear template results into much better and robust ear localization and reduces false positives. The technique in [33] performs ear localization merely based on the size of the connected components which often leads to wrong ear localization as there may exist a cluster of the largest size of non-ear edges.
4. The performance obtained in this technique is found to be robust and stable on a larger data set as compared to [33].

Ear detection results for few profile face images of University of Notre Dame (UND) database are shown in Fig. 2.13 whereas overall localization accuracies for the same database is given in Table 2.3. Ear localization accuracy for UND database

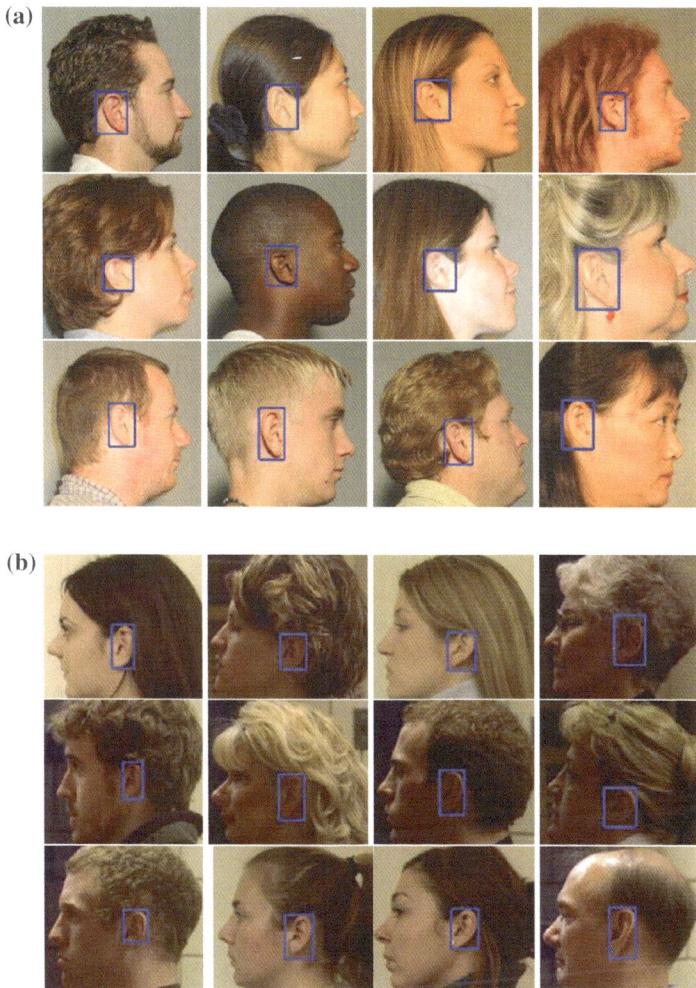

Fig. 2.13 Ear detection results for UND database. **a** UND-E data set, **b** UND-J2 data set

is found to be less as compared to IITK database due to following reason. Hair color of many subjects in UND database is similar to their skin color. Since strength of the discussed technique is derived from the successful detection of skin regions, similarity of the hair color with skin reduces the performance of skin segmentation and in turn, affects the ear localization accuracy and increases false positives.

Table 2.3 also shows comparative performance of some well known techniques on UND database. It is seen from the table that [9] produces low detection rate as compared to the technique discussed in this chapter. Moreover, it makes the assumption that the ear is the principal elliptical shape in the image which limits its use to the controlled environment and frontal ears, as the presence of background objects

Table 2.3 Percentage accuracy for UND database

Technique	Data set	# of test images	Localization accuracy (%)
[9]	Part of UND-J2	942	91
[17]	Part of UND-J2	203	100
Technique discussed here	UND-J2	1780	**96.63**
	UND-E	464	**96.34**

or posed ear may lead to false detections. The technique discussed in [17] achieves good detection rate, but the size of the test data set is very small (only 203 images). Also, if the test ear images are rotated or their appearances are changed with respect to training data, the presented technique may fail because the training images may not include such cases. Forming a database of ears with all possible rotation demands very large space and practically not feasible. Also to detect the ears of different scale, the technique should perform an exhaustive search with filters of various sizes which is computationally very expensive and makes the technique infeasible for real applications. On the other hand, the technique discussed in this chapter can inherently handle rotation (pose) and scale changes and does not incur any extra computational overhead to achieve this. Also, it is tested on a very large data set of 4916 images comprising of rotated (in-plane and out-of-plane) and scaled images which dictates the stability and robustness of the technique. A detailed comparison of [17] with the presented technique is given in Table 2.4.

Table 2.4 Comparison with the technique discussed in [17]

Parameters	Techniques	
	[17]	Presented technique
Time per detection (same configuration)	26.40 s	7.95 s
Training overhead	More. To train classifiers with 1000s of positive and negative samples	Very less. Only required to learn skin parameters and ear template using few 100 samples
Invariant to		
(i) Rotation	No	Yes
(ii) Scale	No	Yes
(iii) Occlusion	Up to some extent	No
Total test data size	Very small (307 images)	Large (4916 images)
Test data	No scaling, minor pose variation	Good amount of scaling and rotation (IITK data sets 2 and 3)

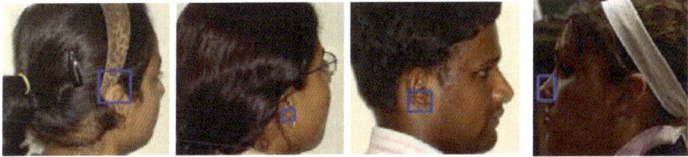

Fig. 2.14 Few failure cases from IITK and UND databases

Performance of the presented technique could not be compared with [4] because of the non-availability of the test results. Also comparisons could not be made with [19, 20] as these techniques have used XM2VTS database [10] which is not available. However, it can be noted that XM2VTS database is relatively easy to work because it contains images captured in plane background with controlled illumination and comprises of good quality images whereas UND images contain non-uniform cluttered background, poor illumination and pose variations.

The presented technique has failed to detects ears fully or partially in some cases of IITK and UND databases. Failure has occurred when ears are occluded by hair or affected by noise and poor illumination. Few examples of failure in detecting ears due to these reasons are shown in Fig. 2.14.

References

1. Burge, Mark and Burger, Wilhelm 2000. Ear biometrics in computer vision. In *Proceedings of International Conference on Pattern Recognition (ICPR'00)*, vol. 2, 822–826.
2. Hurley, David J., Mark S. Nixon, and John N. Carter. 2005. Force field feature extraction for ear biometrics. *Computer Vision and Image Understanding* 98(3): 491–512.
3. Yan, Ping, Kelvin W. Bowyer. 2005. Empirical evaluation of advanced ear biometrics. In *Proceedings of International Conference on Computer Vision and Pattern Recognition-Workshop*, vol. 3, 41–48.
4. Alvarez, L., E. Gonzalez and L. Mazorra. 2005. Fitting ear contour using an ovoid model. In *Proceedings of IEEE International Carnahan Conference on Security Technology (ICCST'05)*, 145–148.
5. Kass, M., A. Witkin, and D. Terzopoulos. 1988. Snakes: Active contour models. *International Journal of Computer Vision* 1(4): 321–331.
6. Yan, Ping, and K.W. Bowyer. 2007. Biometric recognition using 3D ear shape. *IEEE Transactions on Pattern Analysis and Machine Intelligence* 29(8): 1297–1308.
7. Ansari, Saeeduddin and Phalguni Gupta. 2007. Localization of ear using outer helix curve of the ear. In *Proceedings of the International Conference on Computing: Theory and Applications (ICCTA'07)*, 688–692.
8. Yuan, Li, Zhi-Chun Mu. 2007. Ear detection based on skin-color and contour information. In *Proceedings of International Conference on Machine Learning and Cybernetics (ICMLC'07)*, vol. 4, 2213–2217.
9. Arbab-Zavar, Banafshe and Mark S. Nixon. 2007. On shape-mediated enrolment in ear biometrics. In *Proceedings of the 3rd International Conference on Advances in Visual Computing—Volume Part II*, 549–558.

10. Messer, K., J. Matas, J. Kittler, J. Lttin and G. Maitre. 1999. XM2VTSDB: The extended M2VTS database. In *Proceedings of 2nd International Conference on Audio and Video-based Biometric Person Authentication*, 72–77.
11. Sana, Anupam, Phalguni Gupta, Ruma Purkait. 2007. Ear biometric: A new approach. In *Proceedings of International Conference on Advances in Pattern Recognition (ICAPR'07)*, 46–50.
12. Prakash, Surya, Umarani Jayaraman and Phalguni Gupta. 2008. Ear localization from side face images using distance transform and template matching. In *Proceedings of IEEE International Workshop on Image Proceedings Theory, Tools and Applications (IPTA'08)*, 1–8.
13. Prakash, Surya, Umarani Jayaraman and Phalguni Gupta. 2009. A skin-color and template based technique for automatic ear detection. In *Proceedings of International Conference on Advances in Pattern Recognition (ICAPR'09)*, 213–216.
14. Attarchi, S., K. Faez and A. Rafiei. 2008. A new segmentation approach for ear recognition. In *Proceedings of International Conference on Advanced Concepts for Intelligent Vision Systems*, 1030–1037.
15. USTB Database, University of Science and Technology Beijing. http://www.ustb.edu.cn/resb/.
16. Carreira-Perpinan. 1995. Compression neural networks for feature extraction: Application to human recognition from ear images. Master's thesis, Faculty of Informatics, Technical University of Madrid, Spain.
17. Islam, S.M.S., M. Bennamoun and R. Davies. 2008. Fast and fully automatic ear detection using cascaded adaboost. In *Proceedings of IEEE Workshop on Applications of Computer Vision (WACV'08)*, 1–6.
18. Prakash, Surya, Umarani Jayaraman, Phalguni Gupta. 2009. Ear localization using hierarchical clustering. In *Proceedings of SPIE International Defence Security and Sensing Conference, Biometric Technology for Human Identification VI, 730620*, vol. 7306, 730620–730620-9.
19. Cummings, A., M. Nixon and J. Carter. 2010. A novel ray analogy for enrolment of ear biometrics. In *Proceedings of International Conference on Biometrics: Theory, Applications and Systems (BTAS'10)*, 1–6.
20. Ibrahim, Mina I.S., Mark S. Nixon and Sasan Mahmoodi. 2010. Shaped wavelets for curvilinear structures for ear biometrics. In *Proceedings of 6th International Conference on Advances in Visual Computing (ISVC'10)—Part I*, 499–508.
21. Kumar, Amioy, Madasu Hanmandlu, Mohit Kuldeep and H.M. Gupta. 2011. Automatic ear detection for online biometric applications. In *Proceedings of National Conference on Computer Vision, Pattern Recognition, Image Processing and Graphics, NCVPRIPG 2011*, 146–149.
22. Lankton, S., and A. Tannenbaum. 2008. Localizing region-based active contours. *IEEE Transactions on Image Processing* 17(11): 2029–2039.
23. Cai, J., and A. Goshtasby. 1999. Detecting human faces in color images. *Image and Vision Computing* 18(1): 63–75.
24. Poynton, C.A. 1996. *A Technical Introduction to Digital Video*. New York: Wiley.
25. Garcia, C., and G. Tziritas. 1999. Face detection using quantized skin color regions merging and wavelet packet analysis. *IEEE Transactions on Multimedia* 1(3): 264–277.
26. Bay, Herbert, Tinne Tuytelaars and Luc Van Gool. 2006. SURF: Speeded up robust features. In *Proceedings of 9th European Conference on Computer Vision (ECCV'06)*, 404–417.
27. Bay, Herbert, Andreas Ess, Tinne Tuytelaars, and Luc Van Gool. 2008. Speeded-up robust features (SURF). *Computer Vision and Image Understanding* 110(3): 346–359.
28. Lowe, David G. 2004. Distinctive image features from scale-invariant keypoints. *International Journal of Computer Vision* 60(2): 91–110.
29. Mikolajczyk, Krystian, and Cordelia Schmid. 2005. A performance evaluation of local descriptors. *IEEE Transactions on Pattern Analysis and Machine Intelligence* 27(10): 1615–1630.
30. Hopcroft, John, and Robert Tarjan. 1973. Algorithm 447: Efficient algorithms for graph manipulation. *Communications of the ACM* 16(6): 372–378.
31. Bustard, J.D., and M.S. Nixon 2008. Robust 2D ear registration and recognition based on SIFT point matching. In *Proceedings of International Conference on Biometrics: Theory, Applications and Systems (BTAS'08)*, 1–6.

32. University of Notre Dame Profile Face Database, Collections E and J2. http://www.nd.edu/cvrl/CVRL/DataSets.html.
33. Prakash, Surya, Umarani Jayaraman, Phalguni Gupta. 2009. Connected component based technique for automatic ear detection. In: *Proceedings of 16th IEEE International Conference on Image Processing (ICIP'09)*, 2741–2744.

Chapter 3
Ear Recognition in 2D

3.1 Introduction

Most of the well known techniques for 2D ear recognition can be broadly partitioned into following types: appearance based techniques, force field transformation based techniques, geometric features based techniques and neural network based techniques. Appearance based techniques use either global or local appearance of the ear image for recognition. Techniques based on Principal Component Analysis (PCA) [1], Independent Component Analysis (ICA [2]), intensity and color space [3, 4] etc. fall under this category. PCA based technique is an extension of the use of PCA in face recognition. It exploits the training data to find out a set of orthogonal basis vectors representing the directions of maximum variance in the data with minimum reconstruction mean square error. Usually, it drops the first eigenvector assuming that it represents the illumination changes in the image. Victor et al. [5] have used PCA for both face and ear recognition and have concluded that the face performs better than the ear. However, Chang et al. [1] have performed comparison of ear and face recognition and have concluded that ears are essentially just as good as faces for human recognition. This study has reported difference in the rank-1 recognition rates as 71.6 % for the ear and 70.5 % for the face. Chang et al. [1] have suggested that the lower performance of ear in [5] may be due to the reason that the technique in [5] has used images with less control over earrings, hair and lighting. Zhang et al. [2] have used ICA with Radial Basis Function (RBF) network [6] for ear recognition and have shown better performance than PCA when number of features are more. When comparatively less features are available, this study has found that PCA method outperform ICA. However, authors have not dropped the first eigenvector in PCA while comparing the results. Major drawback of the techniques based on PCA or ICA is that they are only usable when images are captured in controlled environment and properly registered. These techniques do not offer any invariance and hence require very accurate registration to achieve consistently good results. Yuizono et al. [7] have modeled the recognition task as an optimization problem and have proposed an ear recognition technique by using a specially developed genetic

© Springer Science+Business Media Singapore 2015
S. Prakash and P. Gupta, *Ear Biometrics in 2D and 3D*,
Augmented Vision and Reality 10, DOI 10.1007/978-981-287-375-0_3

local search targeting the ear images. This technique has reported near 100 % recognition rate on relatively small database of 110 subjects. However, it does not have any invariant properties as it does not include any feature extraction process. Nanni and Lumini [3] have proposed a multi-matcher based technique for ear recognition which exploits appearance based local properties of an ear. It considers overlapping sub-windows to extract local features using bank of Gabor filters [8]. Further, Laplacian Eigen Maps [9] are used to reduce the dimensionality of the feature vectors. Ear is represented using the features obtained from a set of most discriminative sub-windows selected using Sequential Forward Floating Selection (SFFS) algorithm. Matching in this technique is performed by combining the outputs of several 1-nearest neighbor classifiers constructed on different sub-windows. Performance of this technique in terms of Equal Error Rate (*EER*) has been reported as 4.2 % on a database of 114 subjects. Another technique based on fusion of color spaces has been proposed by Nanni and Lumini [4] where few color spaces are selected using SFFS algorithm and Gabor features are extracted from them. Matching is carried out by combining the output of several nearest neighbor classifiers constructed on different color components. Rank-1 recognition performance of this technique is reported to be nearly 84 %.

Force field based techniques [10–12] transform an ear image into a force field and extract features using force field energy functionals discussed in [13]. To transform an image into force field, an image is considered as an array of mutually attracting particles that act as a source of Gaussian force field. Underlying the force field, there exists a scalar potential energy field which, in case of an ear, appears as a smooth surface that looks like a small mountain with a number of peaks joined by ridges. Force field based techniques consider these peaks and ridges as features for ear representation. The directional properties of the force field are utilized to identify the extrema of a small number of potential energy wells and associated potential channels. These "potential wells" and "potential channels" are used as features for ear recognition. The technique has achieved a recognition rate of 99.2 % [11] on XM2VTS face profiles database [14] consisting of 252 images obtained from 63 subjects, 4 samples per subject. This technique has shown much better performance than PCA (which produces recognition rate of only 62.4 % on the same experimental setup) when images are poorly registered. However, when ears are correctly segmented and resized to ear size of 111×73, PCA has produced a recognition rate of 98.4 %, which is due to inherent advantage of correct ear segmentation. Force field based ear recognition has also been found to be robust against noise, adding 18dB of Gaussian noise actually improved the performance to 99.6 % [15]. Mottaleb et al. [16] have also used the force field transform to obtain a smooth surface representation for the ear and then have applied different surface curvature extractors to get the ear features. The technique is tested for identification and has achieved 88 % rank-1 recognition rate for 58 query images of 29 subjects.

Burge and Burger [17, 18] have proposed a technique for ear recognition using geometric information of the ear. The ear has been represented using a neighborhood graph obtained from a Voronoi diagram of the ear edge segments whereas template comparison has been performed through sub-graph matching. Mu et al. [19] have used geometric information of the ear and have reported 85 % recognition rate. In this technique, features are extracted using shape feature vector of the outer ear and the structural feature vector of the inner ear. Choras [20, 21] has used geometric properties of the ear to propose an ear recognition technique in which feature extraction is carried out in two steps. In the first step, global features are extracted whereas the second step extracts local features. Matching is performed in two steps where global features are matched first. Local features are matched only when global features are found to be matching. In another geometry based technique proposed by Shailaja and Gupta [22], an ear is represented by two sets of features, global and local, obtained using outer and inner ear edges respectively. Two ears in this technique are declared similar if they are matched with respect to both the feature sets.

The technique proposed in [23] has treated ear as a planar surface and has created a homography transform using SIFT [24] feature points to register ears accurately. It has achieved results comparable to PCA with manual registration. However, when applied on challenging database, it has shown robustness to background clutter, 20 % occlusion and over $\pm 13°$ of pose variation. In [25], Yuan et al. have proposed a technique for human recognition with partially occluded ear images using neighborhood preserving embedding. Marsico et al. in [26] have proposed a fractal based technique to classify human ears. The technique has adopted feature extraction locally so that the system gets robust with respect to small changes in pose/illumination and partial occlusions. Moreno et al. [27] have carried out ear recognition using two-staged neural network and have reported a recognition accuracy of 93 % on a database of 168 images.

A number of multimodal approaches considering ear with other biometric modalities such as face, speech etc. has also been considered. Iwano et al. [28] have combined ear images with speech using a composite posterior probability. An audio-visual database collected from 38 male speakers in five sessions, acquiring one speech and one ear sample in each session, has been used in experiments. This study has demonstrated the performance improvement over the system using alone either ear or speech. A recognition rate of nearly 91 % has been reported in [1] when a multimodal PCA technique is employed for ear and face on a database consisting of 88 probe and 197 gallery images. Rahman et al. [29] have also proposed multimodal biometric systems using PCA on both face and ear. This study has reported an improved recognition rate of 94.4 % when multimodal biometrics is used instead of individual biometrics on a multimodal database consisting of 90 ear and 90 face images collected from 18 individuals in 5 sessions. Iwano et al. [30] have proposed a multi-algorithmic approach of ear recognition where features from ear images are extracted using PCA and ICA. This study has also shown performance improvement as compared to one when either of the feature extraction is applied alone.

A biometric based security system is expected to fulfill user's demand such as low error rates, high security levels, testing for liveliness of the subject, possibility of fake detection etc. Most of the existing ear recognition techniques have failed to perform satisfactorily in the presence of varying illumination, occlusion and poor image registration. This chapter discusses a new ear based recognition technique which can handle some of these challenges. In this technique, an ear image is enhanced using three image enhancement techniques applied in parallel. SURF feature extractor is used on each enhanced image to extract local features. A multi-matcher system is trained to combine the information extracted from each enhanced image. The technique is found to be robust to illumination changes and performs well even when ear images are not properly registered. The use of multiple image enhancement techniques has made it possible to counteract the effect of illumination, poor contrast and noise while SURF based local feature helps in matching the images which are not properly registered and suffer from pose variations.

The rest of the chapter is organized as follows. Section 3.2 discusses SURF feature extractor and various enhancement techniques used in the ear recognition technique discussed in this chapter. Next section presents the technique for ear recognition. Experimental results of this techniques are analyzed in the last section.

3.2 Preliminaries

Ear images used for recognition may be affected by the problems of contrast, illumination and noise. Hence they are first enhanced before using them for recognition. There are several techniques available in the literature for image contrast enhancement (such as [31–33]), for noise removal (such as [34–36]) and for illumination normalization (such as [37–41]). This section discusses some of the important image enhancement techniques which are used to develop the ear recognition technique discussed here. It briefly presents three image enhancement techniques namely Adaptive Histogram Equalization [33], Non-Local Means Filter [36] and Steerable Gaussian Filter [42] to enhance ear images. Adaptive histogram equalization helps in improving the contrast of the image whereas Non-local means filter reduces the noise in the image. Steerable Gaussian filter helps to reduce the effect of illumination in the image.

3.2.1 Adaptive Histogram Equalization

Contrast limited adaptive histogram equalization (ADHist) [33] can be used to improve the contrast of an image. It divides an image into multiple non-overlapping tiles (regions) and performs histogram equalization for each one individually. This enhances the contrast of each tile. The neighboring tiles are combined together to get the entire enhanced image. ADHist uses bilinear interpolation to remove artificially induced boundaries while combining the tiles. It is capable of limiting the contrast, especially in homogeneous areas, to avoid amplification of any noise that might be present in the image. It improves local contrast of the image and brings out more details in the image.

Let $I \in R^{a \times b}$ be the image of size $a \times b$ to be enhanced. It is divided into the tiles $T_i, i = 1, \ldots, n$ of size $\alpha \times \beta$, for $\alpha < a$ and $\beta < b$ and $n = \lfloor \frac{a \times b}{\alpha \times \beta} \rfloor$. These tiles are enhanced individually and stitched together to get the overall enhanced

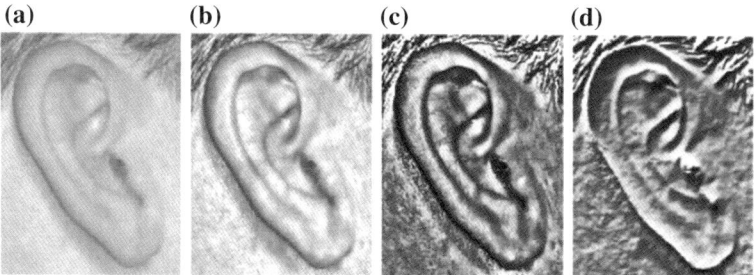

(a) **(b)** **(c)** **(d)**

Fig. 3.1 Image enhancement examples: **a** Original image from UND-E dataset, output after applying, **b** ADHist, **c** NLM and **d** SF enhancement techniques

image I_e. Selection of appropriate values for α and β greatly affects the enhancement performance. These values are found empirically. Consider an image shown in Fig. 3.1a. This image has been enhanced using ADHist and the obtained resultant image is shown in Fig. 3.1b. Steps followed in the enhancement technique are summarized in Algorithm 3.1.

Algorithm 3.1 Enhancement using Adaptive Histogram Equalization

---

- **Input:** Ear image $I \in R^{a \times b}$ of size $a \times b$.
- **Output:** Enhanced image $I_e \in R^{a \times b}$ of size $a \times b$.

1: Divide image I into tiles $T_i, i = 1, \ldots, n$, each of size $\alpha \times \beta$, where $\alpha < a, \beta < b$ and $n = \lfloor \frac{a \times b}{\alpha \times \beta} \rfloor$.
2: **for** $i = 1$ to n **do**
3:    Enhance the contrast of title T_i.
4: **end for**
5: Obtain I_e by stitching all T_is, $i = 1, \ldots, n$. Bilinear interpolation is used to remove all artificial induced boundaries that may occur whiling stitching T_i, $\forall i$ with its neighbors.
6: Return I_e.

3.2.2 Non-local Means Filter

The non-local means (NLM) technique [36] has been proposed for image enhancement by using image denoising. It considers pixel values from the entire image for the task of noise reduction. The algorithm is based on the fact that for every small window of the image, several similar windows can be found in the image and all of these windows can be exploited to denoise the image. Let the noisy image be denoted by $I \in R^{a \times b}$, where a and b are image dimensions and let $p = (x, y) \in I$ stand for an arbitrary pixel location within the noisy image. The NLM algorithm constructs the denoised image I_d by computing each pixel value of I_d as a weighted average of pixels comprising I, i.e. the denoised value $I_d(p)$ at pixel p is given by

$$I_d(p) = \sum_{z \in I} w(p, z) I(z)$$

where $w(z, p)$ represents the weighting function which measures the similarity between the local neighborhoods of the pixels at the spatial locations z and p using Gaussian weighted Euclidian distance. The weighting function used in this equation is defined as follows:

$$w(p, z) = \frac{1}{Z(p)} e^{-\frac{G_\sigma \|I(\Omega_p) - I(\Omega_z))\|_2^2}{h^2}}$$

where,

$$Z(p) = \sum_{z \in I} e^{-\frac{G_\sigma \|I(\Omega_p) - I(\Omega_z))\|_2^2}{h^2}}$$

Here, $Z(p)$ denotes normalizing factor which ensures $\sum_{z \in I} w(p, z) = 1$, G_σ denotes a Gaussian kernel with standard deviation σ, Ω_p and Ω_z are the local neighborhoods of the pixels at locations p and z respectively, h stands for the parameter that controls the decay of the exponential function and $G_\sigma \|.\|_2^2$ denotes Gaussian weighted Euclidian distance. It can be observed that if local neighborhoods of a given pair of pixel locations p and z display a high degree of similarity, the pixels at p and z can be assigned relatively large weights at the time of computing their denoised estimates.

A proper selection of the neighborhood size N and decay parameter h results in a smoothed image with preserved edges. Hence, it can be used to estimate the luminance of an input image and consequently, to compute the (logarithmic) reflectance. An example of the deployment of the NLM algorithm (for a 3×3 local neighborhood and $h = 50$) for estimation of the logarithmic reflectance is shown in Fig. 3.1c. Steps for non-local means filtering are presented in Algorithm 3.2.

3.2.3 Steerable Filter

Steerable Filter (SF) [42] provides an efficient architecture to synthesize filters of arbitrary orientations from linear combinations of basis filters. This allows to adaptively "steer" a filter to any orientation and to determine analytically the filter output as a function of orientation. These filters are normally used for early vision and

---

Algorithm 3.2 Enhancement using Non-local Means Filter

- **Input:** Ear image $I \in R^{a \times b}$ of size $a \times b$.
- **Output:** Enhanced image $I_d \in R^{a \times b}$ of size $a \times b$.

1: **for all** pixels $p = (x, y) \in I, x \in [1, \ldots, a], y \in [1, \ldots, b]$ **do**

2:   $I_d(p) = \sum_{z \in I} w(p, z) I(z)$

 where $w(p, z) = \frac{1}{Z(p)} e^{-\frac{G_\sigma \|I(\Omega_p) - I(\Omega_z)\|_2^2}{h^2}}$, $Z(p) = \sum_{z \in I} e^{-\frac{G_\sigma \|I(\Omega_p) - I(\Omega_z)\|_2^2}{h^2}}$ and $G_\sigma \|.\|_2^2$ denotes

 Gaussian weighted Euclidian distance.

3: **end for**

4: Return I_d.

---

image processing tasks such as angularly adaptive filtering, shape-from-shading, edge detection etc. However, they can also be used to produce illumination invariant representation of an image, such as the gradient image. For example, Gaussian function can be used as the basis filters to obtain steerable filters. To get illumination invariant representation of an image, steerable Gaussian derivatives can be applied at multiple scales and orientations to an image. Enhanced image is computed by taking the weighted linear combination of the filtered images which are obtained by applying the Gaussian derivatives of various scales and orientation to the input image.

There are two critical parameters used to define steerable filters based on Gaussian function: one is σ to define the scale of the filter and another is θ to define the orientation of the filter. To define the Gaussian functions at multiple scales, a set of σ and θ values are required. Let these values be $\{\sigma_1, \sigma_2, \ldots, \sigma_l\}$ and $\{\theta_1, \theta_2, \ldots, \theta_n\}$. Each pair of (σ_i, θ_j) defines one basis filter. Angular spacing of the filters is usually taken equal, hence n values of θ define n angles equally drawn from 0 to 180^0. The choice of values of σ and n depends on the size and content of the image respectively. Input image $I \in R^{a \times b}$ is enhanced by applying all these filters one by one and then by taking the weighted linear combination of outputs of all the filters to produce enhanced image I_s. A weight assigned to a filter is related to importance given to the filter. The ear recognition technique discussed in this chapter considers every filter equally important, hence assigns equal weights to all.

An example of image enhancement of the image shown in Fig. 3.1a using SF technique is given in Fig. 3.1d. It gives an image after normalizing the effect of illumination. Steps for steerable filter based image enhancement are summarized in Algorithm 3.3.

Algorithm 3.3 Enhancement using Steerable Filter

- **Input:** Ear image $I \in R^{a \times b}$ of size $a \times b$.
- **Output:** Enhanced image $I_s \in R^{a \times b}$ of size $a \times b$.

1: Let the set $\{\sigma_1, \sigma_2, \ldots, \sigma_l\}$ and $\{\theta_1, \theta_2, \ldots, \theta_n\}$ define the value of various scales and orientations.
2: Define a set of steerable filters $F_{(1,1)}, F_{(1,2)}, \ldots, F_{(l,n)}$ using Gaussian derivatives where $F_{(i,j)}$
   represents a filter of scale σ_i and orientation θ_j.
3: **for all** pixels $p = (x, y) \in I$ **do**
4:    **for** $\forall (i, j), i \in [1, \ldots, l], j \in [1, \ldots, n]$ **do**
5:       Compute $I_{(i,j)}(p)$ by applying filter $F_{(i,j)}$ at pixel p.
6:    **end for**
7:    Compute $I(p) = \sum_{\forall (i,j)} w_{(i,j)} I_{(i,j)}(p)$ where $w_{(i,j)}$ is the weight assigned to filter $F_{(i,j)}$ and is
      related its importance.
8:    Set $I_s(p) = I(p)$
9: **end for**
10: Return I_s.

3.3 Ear Recognition Technique

This section presents an efficient ear recognition technique. It has made use of three
techniques in parallel to enhance an image producing three different enhanced ver-
sions of the input image. Subsequently, these three feature sets extracted from the
images are used to train three different classifiers. Overview of the system is shown
in Fig. 3.2.

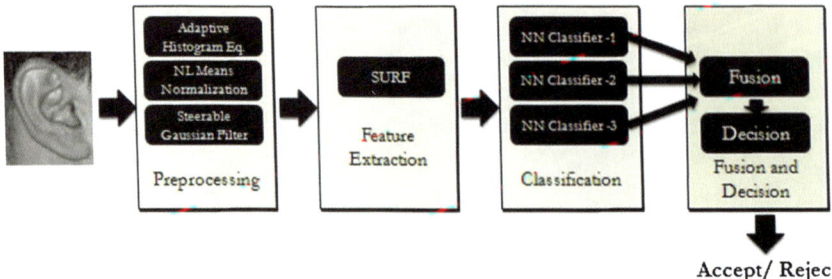

Accept/ Reject

Fig. 3.2 Block diagram of the ear recognition technique

3.3.1 Image Enhancement

This step involves all three image enhancement techniques and is intended to enhance the contrast of the ear image, to minimize the effect of noise and to normalize the effect of illumination and shadow. These techniques have been used in parallel on each input ear image to get three enhanced images. The purpose of image enhancement is to get the accurate SURF feature descriptor vectors for feature points which helps in establishing the correct point correspondence between the feature points in two images. For example, a particular feature point in two different images of the same subject (which are differently illuminated) may get two different SURF descriptor vectors in the absence of enhancement. However, when enhancement is applied on images, descriptor vectors for corresponding points in two images are found to be quite similar.

The use of three enhancement techniques in parallel is based on the following idea. It is observed in real life scenarios that in an environment (in a database), all captured images may not be affected by all issues *viz.* contrast, illumination and noise, together. Depending on the different environmental conditions, they usually suffer by any one of the issue. Hence there is no need to use all the enhancement techniques on an image together (in serial fashion). However, it is ideal to apply all three enhancement techniques in parallel on an image as it is not known a priori the problem by which the image is affected. Hence by using different enhancement techniques in parallel, we can handle all such issues. That means, if an image has the problem of poor contrast, it can be handled by an enhancement technique which can handle contrast problem.

3.3.2 Feature Extraction

Enhanced ear images undergo feature extraction phase where SURF technique (described in Sect. 2.2.2 in detail) is used to extract features. The reason behind selecting SURF for feature extraction over other local feature descriptors such as SIFT [24] and GLOH [43] is as follows. SURF has the ability to capture the properties of spatial localization, change in 3D viewpoint, orientation and scale variations more efficiently as compared to other local descriptors [44]. It provides a highly distinctive descriptor vector for a feature point in the sense that the descriptor vector can be correctly matched with high probability against a large database of descriptor vectors obtained for feature points of many images.

SURF represents an ear image in terms of a set of salient feature points, each point associated with a descriptor vector which can be either of 64 dimensions or 128 dimensions. A 128 dimensional descriptor vector is more discriminative as compared to 64 dimensional vector. Since in ear recognition task, it is always good to represent images with powerful discriminative features, in the this ear recognition technique 128 dimensional descriptor vector is used in feature extraction.

A technique for feature level fusion is used to obtain a fused template for a subject by combining features obtained from multiple training samples (instances) of the subject. If there are n ear images of a subject for training, a fused template for it is obtained by fusing the feature descriptor vectors of all training images together, considering the redundant descriptor vectors only once.

Let SURF feature templates of n training images of a subject be represented by $F_1, F_2, \ldots, F_n$ where F_j is a $128 \times m_j$ dimensional matrix representing feature template for jth ear image with m_j feature points. That is, each column of the matrix F_j represents a descriptor vector of a feature point of jth ear image. Then fused template is essentially obtained by column-wise concatenation of the feature templates considering redundant columns (descriptor vectors) only once. Considering F_js as sets of column vectors, F_{fused} can be mathematically written as follows:

$$F_{fused} = F_1 \cup F_2 \cup \cdots \cup F_n$$

where cardinality of the set F_{fused} provides the number of descriptor vectors in the fused template. Let cardinality of the set F_{fused} be represented by m, then it follows the inequality $m \leq m_1 + m_2 + \cdots + m_n$. Also the size of fused template F_{fused} is $128 \times m$.

Fusion of the templates is done incrementally where first two feature templates F_1 and F_2 are fused to generate a new template T which is further fused with feature template F_3. This procedure is continued until all feature templates are considered for fusion. While fusing two feature templates F_i and F_{i+1}, SURF matching (described in Algorithm 2.1) is applied between the templates to find out the redundant feature descriptor vectors. If a feature descriptor vector in a template matches to a descriptor vector in the another template, it is considered as common to both and is used only once in fusion. Steps of fusion are summarized in Algorithm 3.4

Algorithm 3.4 Computation of Fused SURF Template

- **Input:** A set of SURF feature templates $F_1, F_2, \ldots, F_n$ obtained from n training images of a subject where F_j is a $128 \times m_j$ dimensional matrix representing feature template for jth ear image with m_j feature points.
- **Output:** A fused feature template F_{fused} of size $128 \times m$ where m is number of descriptor vectors in the fused template.

1: Consider F_js as sets of column vectors, each column vector representing a SURF descriptor vector.
2: Set F as NULL.
3: **for** $i = 1$ to n **do**
4:    $F = F \cup F_i$ where $x \in F$ is said to be same as $y \in F_i$ if x and y are matched to each other according to SURF matching.
5: **end for**
6: Return F

3.3.3 Classification and Fusion

Fused feature templates of each subject for various enhancement techniques are used to train nearest neighbor classifiers. Since there are three enhanced techniques used, three sets of fused templates are obtained and thus three nearest neighbor classifiers are trained. Matching strategy used in a nearest neighbor classifier to compare two feature templates is based on SURF matching.

A matching score between two ear feature templates in SURF matching is computed by counting the number of descriptor vector (or their respective feature point) pairs that are matched between the two templates. To get a pair of matching descriptor vectors between two image templates, a descriptor vector from one image template is selected and is compared with all descriptor vectors of the other image template using Euclidean distance. Final matching result is declared using nearest neighbor ratio matching strategy where a pair is said to be matched if its distance is closer than τ times the distance of the second nearest neighbor where τ is the matching threshold. SURF matching steps have been described in Algorithm 2.1.

It is easy to see that in a database, if most of the images are affected by a particular problem, the classifier specific to that problem would perform better for that database. Hence to fuse the scores obtained from different classifiers, relative weights to the classifiers depending upon their individual performance have been assigned. Matching scores obtained from each classifier are normalized using min-max normalization technique and are then fused using weighted sum rule [45]. Final classification decision is taken by using the fused score. Steps followed in training and testing are summarized in Algorithms 3.5 and 3.6 respectively.

Algorithm 3.5 Training of the Ear Recognition System

- **Input:** Training images for M subjects.
- **Output:** Three trained classifiers NN_e, NN_d and NN_s for three enhancement techniques, namely Adaptive Histogram Equalization (ADHist), Non-Local Means (NLM) Filter and Steerable Filter (SF) respectively.

1: Consider set T_i containing n_i training images of ith subject, $i = 1, \ldots, M$
2: /*Enhancement of Training Images*/
3: **for** $i = 1$ to M **do**
4: Compute set T_i^e by enhancing the training images of set T_i using ADHist.
5: Compute set T_i^d by enhancing the training images of set T_i using NLM.
6: Compute set T_i^s by enhancing the training images of set T_i using SF.
7: **end for**
8: /*Training of Classifiers*/
9: Use training set T_i^e, $i = 1, \ldots, M$, to train classifier NN_e
10: Use training set T_i^d, $i = 1, \ldots, M$, to train classifier NN_d
11: Use training set T_i^s, $i = 1, \ldots, M$, to train classifier NN_s
12: Return trained classifiers NN_e, NN_d and NN_s.

Algorithm 3.6 Testing of the Ear Recognition System

- **Input:** Test image I_{test}, claimed identity k, a matching threshold Th, trained classifiers NN_e, NN_d and NN_s and their respective weights A_1, A_2 and A_3.
- **Output:** *Accept* or *Reject* information for the test image.

1: Enhance I_{test} using ADHist, NLM and SF enhancement techniques to obtain enhanced images I_{test}^e, I_{test}^d and I_{test}^s respectively.
2: Compute feature templates F_e^{test}, F_d^{test} and F_s^{test} from enhanced images I_{test}^e, I_{test}^d and I_{test}^s respectively using SURF.
3: Compute similarity score S_1 by matching F_e^{test} with the kth training feature template using NN_e.
4: Compute similarity score S_2 by matching F_d^{test} with the kth training feature template using NN_d.
5: Compute similarity score S_3 by matching F_s^{test} with the kth training feature template using NN_s.
6: Compute weighted fused similarity score $S = \dfrac{A_1 \times S_1 + A_2 \times S_2 + A_3 \times S_3}{A_1 + A_2 + A_3}$
7: **if** $S \geq Th$ **then**
8:    Return *Accept*
9: **else**
10:    Return *Reject*
11: **end if**

3.4 Experimental Results

Experiments are conducted on two databases, namely IIT Kanpur database and University of Notre Dame database (Collections E) [46].

3.4.1 Ear Extraction from the Background

IITK and UND-E databases contain profile face images of human subjects. Ears are segmented from the profile face images using ear segmentation technique discussed in Chap. 2. Manual segmentation is performed for the images where this technique is found to be deficient to segment the ears. Few sample cropped ear images from IITK and UND-E databases are shown in Figs. 3.3 and 3.4 respectively.

3.4.2 Parameters Tuning

Selection of appropriate values of the parameters is critical for achieving the best performance. Parameters which have great impact on the performance are the dimensions of the tiles in ADHist, values of σ and n in SF, values of h and N in NLM and value of τ in SURF Matching.

Since it is difficult to get optimal set of values for these parameters by testing the technique for their all possible values, they are tuned heuristically. To get optimal values of these parameters, a set of 25 subjects is randomly selected from each database and parameter tuning is performed only on each of these data sets. The optimal parameters are used for testing the full database.

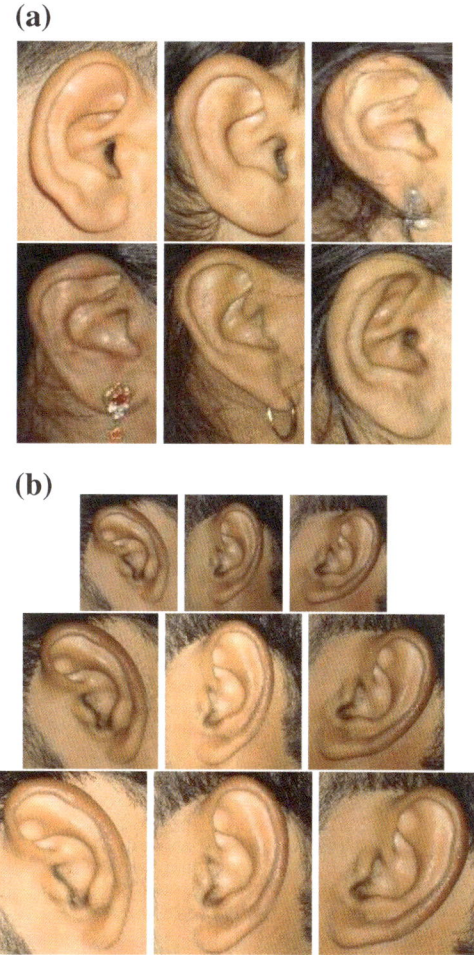

Fig. 3.3 Few sample cropped ear images from IITK database. **a** Data Set 1. **b** Data Set 2

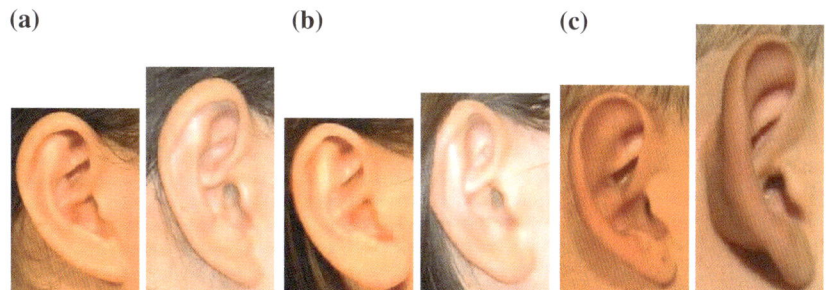

Fig. 3.4 Few sample cropped ear images from UND-E database. **a** Subject ID: 04202. **b** Subject ID: 04217. **c** Subject ID: 04295

Table 3.1 Computation of optimal dimensions of the tile in ADHist for IITK database

Tile size	SURF matching threshold (τ)									
	0.3		0.4		0.5		0.6		0.7	
	EER (%)	EUC (%)	EER (%)	EUC (%)	EER (%)	EUC (%)	EER (%)	EUC (%)	EER (%)	EUC (%)
(a) IITK data Set 1										
2 × 2	3.49	2.58	3.60	1.90	5.47	1.99	8.37	2.85	12.40	5.78
4 × 4	3.51	2.01	3.50	1.25	5.07	1.33	7.89	2.31	12.21	5.41
6 × 6	3.48	1.95	3.50	1.17	4.19	1.10	7.25	2.01	12.23	4.84
8 × 8	3.54	3.03	**3.46**	**1.42**	5.23	1.58	8.44	2.52	12.54	5.36
10 × 10	4.56	4.10	4.15	2.61	5.73	2.23	8.53	3.01	12.25	5.70
12 × 12	8.49	8.62	6.11	5.55	6.64	3.84	9.87	4.55	14.05	7.11
14 × 14	9.91	9.98	5.93	5.23	7.05	4.31	10.31	4.87	14.39	7.46
16 × 16	9.64	9.70	6.40	5.51	7.52	4.48	10.61	4.80	15.11	8.09
18 × 18	8.82	8.77	6.68	5.36	7.02	3.99	10.46	4.84	15.12	8.17
20 × 20	10.63	10.70	7.34	6.27	7.37	4.50	10.56	5.01	15.52	8.23
(b) IITK data Set 2										
2 × 2	2.68	1.76	2.43	1.04	3.87	0.89	7.61	2.12	14.89	6.7
4 × 4	2.66	2.06	**2.25**	**1.03**	4.03	1.07	7.23	2.14	13.89	6.1
6 × 6	3.06	2.64	3.32	1.12	3.84	1.02	6.43	2.01	12.88	5.75
8 × 8	3.42	3.1	4.38	1.93	5.21	1.51	7.58	2.42	14.32	6.39
10 × 10	4.92	4.55	4.27	1.96	5.52	2.01	8.84	2.83	14.97	6.71
12 × 12	7.08	6.73	6.01	3.76	6.51	2.72	9.6	3.8	16.15	7.84
14 × 14	9.64	9.34	6.73	4.63	7.25	2.98	11.46	4.27	17.61	8.84
16 × 16	11.43	11.15	7.47	5.56	8.64	3.71	11.86	4.85	18.87	9.7
18 × 18	12.04	11.83	8.11	6.50	9.77	4.94	12.72	5.82	20.47	10.82
20 × 20	13.33	13.15	9.34	7.99	11.12	6.10	13.54	6.70	21.09	11.12

3.4.2.1 Dimensions of the Tile for ADHist

The technique considers tiles of square size, i.e. $\alpha = \beta$ in the ADHist technique. Dimensions of the tiles are varied from 2 × 2 to 20 × 20 and for each value, Equal Error Rate (*EER*)[1] and Error Under ROC Curve (*EUC*) of the system are computed when only ADHist is used for image enhancement. The tile size which corresponds to minimum *EER* is chosen as the optimal size. Also if two tile sizes give same *EER*, their corresponding *EUC* values are used to break the tie and the tile size for which less *EUC* is obtained, is considered as the optimum tile size. Experiments are conducted to find *EER* and *EUC* for IITK and UND-E databases which are shown in Tables 3.1 and 3.2 respectively. It can be observed from the tables that the optimal values of tile size for IITK database Set 1 and Set 2 are 8 × 8 and 4 × 4 respectively while that for UND-E database is 16 × 16.

---

[1] Various performance measures including *EER* and *EUC* are explained in Chap. 1.

Table 3.2 Computation of optimal dimensions of the tile in ADHist for UND-E database

Tile size	SURF matching threshold (τ)									
	0.3		0.4		0.5		0.6		0.7	
	EER (%)	EUC (%)	EER (%)	EUC (%)	EER (%)	EUC (%)	EER (%)	EUC (%)	EER (%)	EUC (%)
2×2	11.75	6.73	9.97	3.93	8.83	3.53	9.47	3.63	11.44	4.4
4×4	15.14	8.37	10.42	4.71	10.11	3.12	9.57	3.22	11.5	4.08
6×6	13.27	7.05	9.62	3.9	9.47	3.71	9.78	3.53	10.73	3.94
8×8	13.2	7.93	9.34	4.21	8.68	3.04	8.14	2.48	9.63	3.26
10×10	13.88	9.33	9.66	4.19	8.1	2.71	8.46	2.60	10.00	3.38
12×12	12.81	8.48	10.48	4.82	8.20	2.53	7.24	2.18	8.03	2.89
14×14	13.00	9.24	10.78	4.87	8.31	3.16	8.06	2.74	8.15	2.36
16×16	12.98	9.55	10.51	3.99	8.07	2.88	**6.72**	**2.40**	8.39	2.41
18×18	17.91	10.69	13.05	5.16	10.14	3.26	8.2	2.67	8.67	2.65
20×20	13.2	10.01	10.32	4.02	7.96	2.57	7.33	2.12	7.26	2.22

We have noticed that the changes in *EER* and *EUC* are gradual in Tables 3.1 and 3.2 except a few exceptions. In Table 3.1, values of *EER* and *EUC* are gradually increased. However if one observes Table 3.2, it is found that for $\tau = 0.6$ and $\tau = 0.7$, *EER* and *EUC* are gradually decreased while for $\tau = 0.4$ and $\tau = 0.5$, *EER* and *EUC* are almost consistent. For $\tau = 0.3$, its behaviour is little abrupt because SURF matching at low threshold is not very stable. Also this may be due to the illumination and contrast variations in UND-E data set. But this is not the case with IITK data sets.

From Table 3.1a, it can be seen that error values are almost same for $\tau = 0.7$ and for different tile sizes lying between 2×2 and 10×10 or that between 12×12 and 14×14 or that between 16×16 and 20×20. Thus, little change in the tile size does not significantly change the error values.

3.4.2.2 Values of *h* and *N* in NLM Filters

In NLM filters, h controls the decay of the exponential function while N is the neighborhood size (i.e., the size of the patches to be used in the NLM algorithm). To search the optimal values, h and N are changed between 20 to 100 and 4 to 6 respectively. For each combination of (h, N), image enhancement is performed and the enhanced image is used for recognition. The values of *EER* and *EUC* of the system are computed and (h, N) pair which corresponds to minimum *EER* is considered as optimal. Further, *EUC* is used to break the tie in case of two or more (h, N) pairs give same *EER*. Experiments are conducted to find *EER* and *EUC* for IITK and UND-E databases which are shown in Tables 3.3 and 3.4 respectively. It is observed that the optimal values of (h, N) for IITK database Set 1 and Set 2 are $(100, 6)$ and $(50, 6)$ respectively while for UND-E database it is $(50, 6)$.

Table 3.3 Computation of optimal values of h and N in NLM filters for IITK database

h	N	SURF matching threshold (τ)									
		0.3		0.4		0.5		0.6		0.7	
		EER (%)	EUC (%)	EER (%)	EUC (%)	EER (%)	EUC (%)	EER (%)	EUC (%)	EER (%)	EUC (%)
(a) IITK data Set 1											
20	4	11.31	11.42	5.63	5.63	4.52	3.85	5.64	2.95	8.99	3.86
	5	16.60	16.79	9.78	9.90	6.31	6.20	6.54	4.90	9.34	4.53
	6	29.21	29.48	20.17	20.50	14.34	14.67	11.92	12.13	11.16	9.94
50	4	5.16	5.01	3.97	2.86	5.21	1.97	7.61	2.34	11.47	4.81
	5	5.33	5.28	4.31	3.32	4.30	1.92	6.51	2.18	10.98	4.51
	6	5.33	5.33	3.80	3.08	4.36	2.37	6.27	2.21	10.77	4.21
80	4	3.36	2.77	2.99	1.32	4.52	1.07	7.79	2.06	12.64	4.98
	5	3.46	2.93	3.25	1.51	4.80	1.07	7.69	2.05	12.32	4.99
	6	4.95	4.75	3.30	2.21	4.36	1.54	7.34	2.26	12.11	4.93
100	4	3.48	2.82	3.12	1.43	4.94	1.30	7.54	2.18	12.83	5.26
	5	3.40	2.80	3.03	1.44	4.77	1.25	7.93	2.10	12.76	5.15
	6	3.54	2.99	**2.90**	**1.17**	4.78	1.14	7.57	2.01	12.24	4.93
(b) IITK data Set 2											
20	4	22.01	22.01	11.41	11.37	6.38	5.92	6.87	4.01	10.52	4.39
	5	28.05	28.05	17.41	17.40	10.42	10.35	7.64	6.75	11.40	5.79
	6	34.10	34.10	21.88	21.88	13.46	13.43	8.64	8.20	11.13	7.03
50	4	4.35	4.04	4.41	2.37	5.24	1.26	8.00	2.38	13.99	5.91
	5	5.02	4.84	3.90	2.24	4.76	1.41	7.29	1.98	12.74	5.31
	6	5.76	5.67	**3.48**	**2.32**	4.49	1.12	6.85	1.86	12.29	4.83
80	4	3.72	3.27	3.98	1.60	4.40	1.34	7.31	2.30	14.45	6.11
	5	3.88	3.46	4.08	1.67	4.42	0.94	7.37	2.12	14.23	6.01
	6	3.95	3.59	3.95	1.80	4.35	1.19	7.44	2.08	13.71	5.78
100	4	3.86	3.36	4.11	1.62	4.26	1.33	7.63	2.18	14.07	6.05
	5	3.89	3.44	3.71	1.54	4.64	1.29	7.35	2.14	14.01	6.07
	6	3.74	3.30	4.08	1.73	4.49	1.26	7.32	2.25	13.60	5.89

3.4.2.3 Values of σ and n in SF

In steerable filters, σ defines a vector of length l where l is the number of filter scales and n is the angular resolution of filters. In our experiments, we have considered two sets of σ: $\{0.5, 1, 1.5, 2, 2.5\}$ and $\{0.1, 1, 2, 3, 4\}$ while value of n is taken as 4 (i.e., $\theta = 0, \frac{\pi}{4}, \frac{\pi}{2}, \frac{3\pi}{4}$), 6 (i.e., $0, \frac{\pi}{6}, \frac{\pi}{3}, \frac{\pi}{2}, \frac{2\pi}{3}, \frac{5\pi}{6}$) and 8 (i.e., $0, \frac{\pi}{8}, \frac{\pi}{4}, \frac{3\pi}{8}, \frac{\pi}{2}, \frac{5\pi}{8}, \frac{3\pi}{4}, \frac{7\pi}{8}$). For each combination of σ and n, image enhancement is performed and enhanced image is used for recognition using SURF features and nearest neighbor classifier. The values of *EER* and *EUC* of the system are computed and (σ, n) pair which corresponds to minimum *EER* is considered as

Table 3.4 Computation of optimal values of h and N in NLM filters for UND-E database

h	N	SURF matching threshold (τ)									
		0.3		0.4		0.5		0.6		0.7	
		EER (%)	EUC (%)	EER (%)	EUC (%)	EER (%)	EUC (%)	EER (%)	EUC (%)	EER (%)	EUC (%)
20	4	20.85	20.84	11.07	10.82	9.11	6.65	8.02	3.1	7.12	2.27
	5	21.75	21.74	11.68	11.61	8.54	7.51	7.63	4.29	7.77	2.84
	6	27.05	27.05	13.78	13.75	9.00	8.58	8.67	4.40	7.77	2.72
50	4	10.55	10.07	9.85	5.14	7.31	2.63	5.80	1.83	5.97	1.76
	5	11.77	11.58	8.96	6.29	7.52	2.92	6.22	1.90	5.79	1.55
	6	13.42	13.34	8.62	7.21	6.59	3.7	5.79	1.81	**5.75**	**1.40**
80	4	9.47	7.44	8.22	3.30	7.07	2.54	5.87	1.62	5.89	1.64
	5	12.46	9.4	10.09	4.62	7.85	2.5	5.84	1.86	5.89	1.58
	6	12.81	10.13	9.63	4.75	7.27	2.53	5.85	1.48	5.83	1.33
100	4	12.71	7.86	10.87	4.46	8.49	2.88	5.89	1.78	6.21	1.68
	5	10.31	8.03	8.09	3.26	6.83	2.30	5.99	1.94	5.90	1.56
	6	10.91	9.25	7.31	3.87	6.13	2.35	5.85	1.47	5.85	1.50

optimal. Various *EER* and *EUC* values for IITK and UND-E databases are shown in Table 3.5. It is observed from the table that the optimal values of parameters (σ, n) for SF are ({0.5, 1, 1.5, 2, 2.5}, 8) and ({0.1, 1, 2, 3, 4}, 8) for IITK database Set 1 and Set 2 respectively while ({0.5, 1, 1.5, 2, 2.5}, 8) for UND-E database. Further, there are two values of (σ, n) pair (i.e., ({0.1, 1, 2, 3, 4}, 6) and ({0.1, 1, 2, 3, 4}, 8)) in Table 3.5b for which *EER* attains minimum value. To break such type of tie, *EUC* is used and ({0.1, 1, 2, 3, 4}, 8) is chosen as the optimal parameter set as it has minimum *EUC* value among the two.

3.4.2.4 Value of τ for SURF Matching

A match between two descriptor vectors in SURF matching is determined by computing the ratio of distances from the closest neighbor to the distance of the second closest neighbor. All matches in which distance ratio is greater than τ are rejected. Experiments are performed on a selected set of images of IITK and UND-E databases by changing the value of τ from 0.3 to 0.7 with an increment of 0.1. This range of values is used in each of the experiment conducted to determine the parameters of ADHist, NLM and SF filters. For each enhancement technique, a value of τ is determined. Values of all these parameters discussed above are summarized in Table 3.6.

Table 3.5 Computation of optimal values of σ and n in SF for IITK and UND-E databases

σ	n	SURF matching threshold (τ)									
		0.3		0.4		0.5		0.6		0.7	
		EER (%)	EUC (%)	EER (%)	EUC (%)	EER (%)	EUC (%)	EER (%)	EUC (%)	EER (%)	EUC (%)
(a) IITK data Set 1											
{0.5, 1, 1.5, 2, 2.5}	4	3.50	1.76	3.52	1.27	4.04	1.34	6.67	1.99	11.45	4.46
	6	3.48	2.95	3.60	2.07	5.23	2.02	7.17	2.59	12.44	5.16
	8	3.50	1.63	**3.46**	**0.98**	4.85	1.32	7.36	2.25	12.47	5.19
{0.1, 1, 2, 3, 4}	4	3.51	2.07	3.52	1.52	4.28	1.51	6.83	2.10	12.11	5.02
	6	3.53	2.02	3.60	1.42	4.94	1.64	7.76	2.51	12.61	5.45
	8	3.56	1.84	3.58	1.25	5.02	1.46	7.31	2.46	12.97	5.75
(b) IITK data Set 2											
{0.5, 1, 1.5, 2, 2.5}	4	3.83	3.02	4.25	1.63	4.77	1.54	7.43	2.68	13.57	6.07
	6	4.11	3.23	3.86	1.54	5.01	1.49	7.34	2.68	13.77	6.17
	8	3.30	2.16	4.00	1.19	5.02	1.34	8.02	2.61	14.33	6.47
{0.1, 1, 2, 3, 4}	4	4.26	3.42	3.74	1.35	4.51	1.48	7.96	2.86	14.60	6.68
	6	3.60	2.76	3.28	1.15	4.77	1.34	7.99	2.64	15.06	6.83
	8	3.45	2.61	**3.28**	**1.11**	4.58	1.29	8.00	2.75	14.21	6.59
(c) UND-E database											
{0.5, 1, 1.5, 2, 2.5}	4	12.92	9.69	10.64	5.05	9.28	3.62	8.92	3.17	8.18	2.72
	6	12.27	6.96	10.04	3.53	7.41	2.65	6.83	2.25	6.53	2.13
	8	12.41	7.37	9.15	3.5	7.1	1.79	**6.51**	**1.67**	6.61	1.72
{0.1, 1, 2, 3, 4}	4	13.71	7.47	10.68	4.66	9.83	3.9	9.24	3.41	9.66	4.07
	6	13.39	7.55	10.37	4.29	8.75	3.06	8.02	2.71	8.23	2.8
	8	14.45	7.62	10.45	4.6	8.01	2.73	8.02	2.48	7.62	2.63

Table 3.6 Optimal parameters for the ear recognition technique

Enhancement technique	Parameter	Databases		
		IITK Set 1	IITK Set 2	UND-E
ADHist	Tile size	8×8	4×4	16×16
	τ	0.4	0.4	0.6
NLM	h	100	50	50
	N	6	6	6
	τ	0.4	0.4	0.7
SF	σ	{0.5, 1, 1.5, 2, 2.5}	{0.1, 1, 2, 3, 4}	{0.5, 1, 1.5, 2, 2.5}
	n	8	8	8
	τ	0.4	0.4	0.6

3.4.3 Results

The performance of a biometric system can be measured in terms of recognition accuracy, equal error rate (*EER*) and error under *ROC* curves (*EUC*). Values of recognition accuracy (with corresponding *FAR* and *FRR*), *EER*, *EUC* for IITK ear database for various combinations of enhancement techniques are given in Tables 3.7 and 3.8. It can be observed that the best results are obtained when all three image enhancement techniques are employed in the recognition process. *ROC* curves for IITK database Set 1 and Set 2 are shown in Fig. 3.5. The *ROC* curves obtained for the technique employing all three image enhancement techniques is found to be superior to others.

Accuracy obtained in Table 3.8 is always greater than that shown in Table 3.7 except for NLM. Greater accuracy in Table 3.8 is achieved due to the fact that in IITK database Set 2, all subjects are having 9 samples while in Set 1, number of samples varies from 2 to 10 (almost 50 % subjects have number of samples less than 4). This provides better training in Set 2 compared to Set 1 which leads to better accuracy.

Table 3.7 Performance on IITK data Set 1 for various combinations of the enhancement techniques

Enhancement techniques	Accuracy (*FAR,FRR*)	*EER*	*EUC*
No enhancement	93.93(5.71,6.42)	6.15	2.52
ADHist	96.54(2.89,4.04)	3.46	1.42
NLM	97.10(3.07,2.72)	2.90	1.17
SF	96.68(2.92,3.72)	3.46	0.98
ADHist + NLM	97.25(2.83,2.67)	2.98	0.90
ADHist + SF	97.13(2.92,2.82)	3.09	0.80
NLM + SF	97.20(2.71,2.89)	2.94	0.83
ADHist + NLM + SF	**97.35(2.70,2.60)**	**2.88**	**0.75**

Table 3.8 Performance on IITK data Set 2 for various combinations of the enhancement techniques

Enhancement techniques	Accuracy (*FAR,FRR*)	*EER*	*EUC*
No enhancement	94.56(5.11,5.77)	5.57	1.62
ADHist	97.94(1.42,2.70)	2.25	1.03
NLM	96.55(2.10,4.79)	3.48	2.32
SF	96.85(1.70,4.61)	3.28	1.11
ADHist + NLM	98.17(1.49,2.17)	2.11	0.58
ADHist + SF	98.62(1.07,1.69)	1.68	0.40
NLM + SF	98.07(1.83,2.02)	2.26	0.48
ADHist + NLM + SF	**98.79(0.88,1.54)**	**1.59**	**0.36**

Fig. 3.5 *ROC* curves for
IITK data sets showing the
performance for various
combinations of
enhancement techniques.
a Data Set 1. **b** Data Set 2

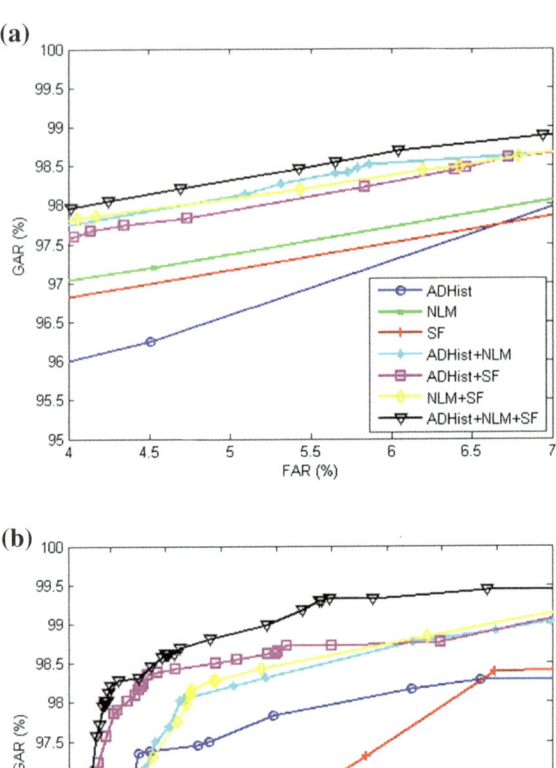

Table 3.9 gives the values of various performance measures for UND-E database
for various combinations of enhancement techniques. It has been noticed that the
best results are obtained when all three enhancement techniques are employed in
recognition process. The best *EER* and *EUC* are found to be much less than those
reported in two well known ear recognition techniques ([3, 4]). Comparative performance of the discussed ear recognition technique with the best known results for
UND-E database is summarized in Table 3.10 All results presented here are averaged over 30 experiments; hence they show more stable performance compared to
the results reported in [3, 4] where they are averaged only for 10 and 20 experiments
respectively. *ROC* curves for UND-E database are shown in Fig. 3.6 where the *ROC*
curve employing all three image enhancement techniques is found to be superior
to others.

Table 3.9 Performance on UND-E database for various combinations of the enhancement techniques

Enhancement techniques	Accuracy (*FAR,FRR*)	*EER*	*EUC*
No enhancement	90.05(6.71,13.19)	10.64	3.96
ADHist	93.64(5.18,7.54)	6.72	2.40
NLM	95.25(2.31,7.19)	5.75	1.40
SF	94.17(3.31,8.36)	6.51	1.67
ADHist + NLM	96.13(2.97,4.77)	4.40	1.34
ADHist + SF	95.41(4.01,5.18)	5.06	1.49
NLM + SF	96.31(2.85,4.53)	4.22	**1.13**
ADHist + NLM + SF	**96.75(2.58,3.92)**	**3.80**	1.16

Table 3.10 Comparison of performance with the latest reported results for UND-E database

Technique	Accuracy	*EER* (*FAR,FRR*)	*EUC*
Proposed in [3]	–	4.20	3.00[a]
Proposed in [4]	–	–	1.50
Technique discussed here	**96.75(2.58,3.92)**	**3.80**	**1.13**

[a]Reported in [4] for the technique proposed in [3]

Fig. 3.6 *ROC* curves for IITK data sets showing the performance for various combinations of enhancement techniques

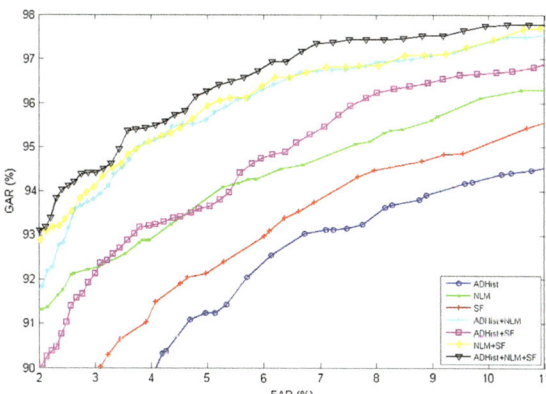

Score level fusion is performed by using weighted sum rule [45]. In this ear recognition technique, relative weights for different classifiers are learnt by using a set of images randomly selected from the database and then by performing recognition task on them. Let the independent use of classifiers C_1, C_2 and C_3 produces classification accuracies of A_1, A_2 and A_3 respectively. In this technique, these accuracies are used as the weight of individual classifiers for fusion. The fusion score is given by

$$S = \frac{A_1 \times S_1 + A_2 \times S_2 + A_3 \times S_3}{A_1 + A_2 + A_3}$$

where S_1, S_2, S_3 are the individual scores produced by classifiers C_1, C_2 and C_3 respectively. *ROC* curves, shown in Figs. 3.5 and 3.6, are drawn for the system which use the weighted sum rule for fusion of matching scores obtained through three classifiers.

References

1. Chang, Kyong, Kevin W. Bowyer, Sudeep Sarkar, and Barnabas Victor. 2003. Comparison and combination of ear and face images in appearance-based biometrics. *IEEE Transactions on Pattern Analysis and Machine Intelligence* 25(9): 1160–1165.
2. Zhang, H.J., Z.C. Mu, W. Qu, L.M. Liu, and C.Y. Zhang. 2005. A novel approach for ear recognition based on ICA and RBF network. In *Proceedings of 4th International Conference on Machine Learning and Cybernetics (CMLC'05)*, 4511–4515.
3. Nanni, Loris, and Alessandra Lumini. 2007. A multi-matcher for ear authentication. *Pattern Recognition Letters* 28(16): 2219–2226.
4. Nanni, Loris, and Alessandra Lumini. 2009. Fusion of color spaces for ear authentication. *Pattern Recognition* 42(9): 1906–1913.
5. Victor, Barnabas, Kevin Bowyer, and Sudeep Sarkar. 2002. An evaluation of face and ear biometrics. In *Proceedings of 16th International Conference on Pattern Recognition (ICPR'02)*, vol. 1, 429–432.
6. Buhmann, Martin D., and M.D. Buhmann. 2003. *Radial basis functions*. New York: Cambridge University Press.
7. Yuizono, T., Y. Wang, K. Satoh, and S. Nakayama. 2002. Study on individual recognition for ear images by using genetic local search. In *Proceedings of 2002 Congress on Evolutionary Computation, 2002 (CEC'02)*, vol. 1, 237–242.
8. Feichtinger, Hans G., and T. Strohmer (eds.). 1997. *Gabor analysis and algorithms: theory and applications*, 1st ed. Boston: Birkhauser.
9. Belkin, Mikhail, and Partha Niyogi. 2003. Laplacian eigenmaps for dimensionality reduction and data representation. *Neural Computation* 15(6): 1373–1396.
10. Hurley, David J., Mark S. Nixon, and John N. Carter. 2000. Automatic ear recognition by force field transformations. In *Proceedings of IEE Colloquium: Visual Biometrics*, 7/1–7/5.
11. Hurley, David J., Mark S. Nixon, and John N. Carter. 2005. Force field feature extraction for ear biometrics. *Computer Vision and Image Understanding* 98(3): 491–512.
12. Hurley, D.J., M.S. Nixon, and J.N. Carter. 2000. A new force field transform for ear and face recognition. In *Proceedings of International Conference on Image Processing (ICIP'00)*, vol. 1, 25–28.
13. Hurley, D., M. Nixon, and J. Carter. 2002. Force field energy functionals for image feature extraction. *Image and Vision Computing* 20(5–6): 311–317.
14. Messer, K., J. Matas, J. Kittler, J. Lttin, and G. Maitre. 1999. XM2VTSDB: The extended M2VTS database. In *Proceedings of 2nd International Conference on Audio and Video-based Biometric Person Authentication*, 72–77.
15. Hurley, D.J., M.S. Nixon, and J.N. Carter. 2005. Ear biometrics by force field convergence. In *Proceedings of International Conference on Audio- and Video-based Biometric Person Authentication*. LNCS, vol, 3546, 119–128.
16. Abdel-Mottaleb, Mohamed, and J.D. Zhou. 2006. Human ear recognition from face profile images. In *Proceedings of International Conference on Biometrics (ICB 2006)*, 786–792.
17. Burge, M., and W. Burger. 1997. Ear biometrics for machine vision. In *Proceedings of 21st workshop of the Austrian Association for Pattern Recognition (WAAPR'97)*, Hallstatt.
18. Burge, Mark, and Wilhelm Burger. 2000. Ear biometrics in computer vision. In *Proceedings of International Conference on Pattern Recognition (ICPR'00)*, vol. 2, 822–826.

19. Mu, Z., L. Yuan, Z. Xu, D. Xi, and S. Qi. 2004. Shape and structural feature based ear recognition. In *Proceedings of Advances in Biometric Person Authentication*. LNCS, vol. 3338, 663–670.

20. Choras, Michal. 2005. Ear biometrics based on geometrical feature extraction. *Electronic Letters on Computer Vision and image Analysis* 5(3): 84–95.

21. Choras, Michal. 2006. Further developments in geometrical algorithms for ear biometrics. In *Proceedings of 4th International Conference on Articulated Motion and Deformable Objects (AMDO'06)*, 58–67.

22. Shailaja, Dasari, and Phalguni Gupta. 2006. A simple geometric approach for ear recognition. In *Proceedings of 9th International Conference on Information Technology (ICIT'06)*, 164–167.

23. Bustard, J.D., and M.S. Nixon. 2008. Robust 2D ear registration and recognition based on SIFT point matching. In *Proceedings of International Conference on Biometrics: Theory, Applications and Systems (BTAS'08)*, 1–6.

24. Lowe, David G. 2004. Distinctive image features from scale-invariant keypoints. *International Journal of Computer Vision* 60(2): 91–110.

25. Yuan, Li, Zhen-Hua Wang, and Zhi-Chun Mu. 2010. Ear recognition under partial occlusion based on neighborhood preserving embedding. In *Proceedings of SPIE International Defence Security and Sensing Conference, Biometric Technology for Human Identification VII*, vol. 7667, 76670b–76670b-13.

26. De Marsico, M., N. Michele, and D. Riccio. 2010. HERO: Human ear recognition against occlusions. In *Proceedings of International Conference on Computer Vision and Pattern Recognition-Workshop*, 178–183.

27. Moreno, B., A. Sanchez, and J.F. Velez. 1999. On the use of outer ear images for personal identification in security applications. In *Proceedings of International Carnahan Conference on Security Technology*, 469–476.

28. Iwano, K., T. Hirose, E. Kamibayashi, and S. Furui. 2003. Audio-visual person authentication using speech and ear images. In *Proceedings of Workshop on Multimodal User Authentication*, 85–90.

29. Rahman, M.M., and S. Ishikawa. 2005. Proposing a passive biometric system for robotic vision. In *Proceedings of 10th International Symposium on Artificial Life and Robotics (AROB'05)*, 4–6.

30. Iwano, K., T. Miyazaki, and S. Furui. 2005. Multimodal speaker verification using ear image features extracted by pca and ica. In *Proceedings of International Conference on Audio and Video Based Biometric Person Authentication*. LNCS, vol. 3546, 588–5996.

31. Agaian, Sos S., B. Silver, and K.A. Panetta. 2007. Transform coefficient histogram-based image enhancement algorithms using contrast entropy. *IEEE Transactions on Image Processing* 16(3): 741–758.

32. Silver, Blair, Sos S. Agaian, and Karen Panetta. 2005. Logarithmic transform coefficient histogram matching with spatial equalization. In *Proceedings of SPIE 5817, Visual Information Processing XIV*, 237.

33. Zuiderveld, Karel. 1994. Contrast limited adaptive histogram equalization. *Graphics gems IV*, 474–485. San Diego: Academic Press Professional Inc.

34. Agaian, Sos S., Karen P. Lentz, and Artyom M. Grigoryan. 2000. A new measure of image enhancement. In *Proceedings of IASTED International Conference on Signal Processing and Communication*.

35. Kogan, Robert G., Sos S. Agaian, and Karen Panetta Lentz. 1998. Visualization using rational morphology and zonal magnitude reduction. In *Proceedings of SPIE 3304, Nonlinear Image Processing IX*, vol. 153.

36. Štruc, Vitomir, and Nikola Pavešić. 2009. Illumination invariant face recognition by non-local smoothing. In *Proceedings of Joint COST 2101 and 2102 iNternational Conference on Biometric ID Management and Multimodal Communication (BioID MultiComm'09)*. LNCS, vol. 5707, 1–8.

37. Agaian, Sos S. 1999.

38. Agaian, Sos S., K. Panetta, and A.M. Grigoryan. 2001. Transform-based image enhancement algorithms with performance measure. *IEEE Transactions on Image Processing* 10(3): 367–382.
39. Silva, Eric A., Karen Panetta, and Sos S. Agaian. 2007. Quantifying image similarity using measure of enhancement by entropy. In *Proceedings of SPIE 6579, Mobile Multimedia/Image Processing for Military and Security Applications 2007*, 65790U.
40. Wharton, Eric, Sos S. Agaian, and Karen Panetta. 2006. Comparative study of logarithmic enhancement algorithms with performance measure.
41. Wharton, Eric, Sos S. Agaian, and Karen Panetta. 2006. A logarithmic measure of image enhancement. In *Proceedings of SPIE Vol. 6250, Mobile Multimedia/Image Processing for Military and Security Applications*, 62500P.
42. Freeman, William T., and Edward H. Adelson. 1991. The design and use of steerable filters. *IEEE Transactions on Pattern Analysis and Machine Intelligence* 13(9): 891–906.
43. Mikolajczyk, Krystian, and Cordelia Schmid. 2005. A performance evaluation of local descriptors. *IEEE Transactions on Pattern Analysis and Machine Intelligence* 27(10): 1615–1630.
44. Bay, Herbert, Andreas Ess, Tinne Tuytelaars, and Luc Van Gool. 2008. Speeded-up robust features (SURF). *Computer Vision and Image Understanding* 110(3): 346–359.
45. Jayaraman, Umarani, Surya Prakash, and Phalguni Gupta. 2008. Indexing multimodal biometric databases using Kd-tree with feature level fusion. In *Proceedings of 4th International Conference on iNformation Systems Security (ICISS'08)*. LNCS, vol. 5352, 221–234.
46. University of Notre Dame Profile Face Database, Collection E. http://www.nd.edu/cvrl/CVRL/DataSets.html.

Chapter 4
Ear Detection in 3D

4.1 Introduction

Similar to 2D, ear recognition in 3D also involves two major steps viz. (i) detection and segmentation of ear from the profile face and (ii) recognition using segmented ear. Most of available well known ear recognition techniques have directly focussed on recognition phase by making use of manually cropped ears. There exist a few techniques which automatically crop ear from 3D profile face range images and use for recognition. In [1], an ear detection technique in 3D using ear template has been proposed. It represents model template by an averaged histogram of shape index of the ear while ear detection is carried out by performing template matching at potential areas in the profile face image. On a database of 30 subjects with two images of each, the technique has produced 91.5 % correction detection rate. In [2], an ear has been represented by a set of discrete 3D vertices computed from helix and antihelix edges of the ear. Step edges obtained from the range image are clustered to detect the ear. Each edge cluster is registered with the ear template and the one having minimum mean registration error is claimed as the ear. The technique has been tested on a database of 32 subjects, each subject having 6 images. It has produced 92.6 % correct ear detection rate. The performance has been enhanced in [3] where a single reference 3D ear shape model is used to locate ear helix and antihelix curves of the ear. However, this technique also makes use of a registered 2D color image together with 3D range image for ear detection. It has been tested on two databases *viz.* database of University of California Riverside (UCR) and Collections F and a subset of Collection G database of University of Notre Dame (UND). On UCR database with 902 profile face range images collected from 155 subjects, the technique has achieved 99.3 % percent correct detection rate while 87.71 % correct detection rate has been found on UND database of 700 images collected from 326 subjects (302 subjects, 302×2 images of Collection F and 24 subjects, 24×4 images from Collection G). All these techniques are template based and hence are not efficient for handling scale and pose variations in the range data.

© Springer Science+Business Media Singapore 2015
S. Prakash and P. Gupta, *Ear Biometrics in 2D and 3D*,
Augmented Vision and Reality 10, DOI 10.1007/978-981-287-375-0_4

In [4], a manual technique for segmenting ears using landmark points *viz.* Triangular Fossa and Incisure Intertragica on the 3D profile range image has been proposed. A line is drawn by joining these landmark points to obtain the orientation and scaling of the ear. This information has been used in rotating and scaling a mask which is applied on the original 3D range image to crop the ear data. In [5], there exists another 3D ear detection technique in range images using 2D registered image of profile face along with 3D. The technique proposed in [5] uses a part of UND Collection J2 (total test images 415) database for experiments. It has achieved ear detection accuracy of 85.54 % when it uses only 3D information and has reported to achieve 100 % detection performance when information from both 2D and 3D is used for 3D ear detection. The technique locates nose tip to obtain the probable ear region and finds the ear pit by using skin and curvature information in this region. An active contour is initialized using the boundary of the pit and both color and depth information are utilized to converge the contour at the ear boundary. Its performance should be declined when profile face is affected by pose variations.

In [6], Zhou et al. have introduced a shape-based feature set, termed as Histograms of Categorized Shapes (HCS), for robust 3D object recognition and have used it for 3D ear detection. The technique has achieved 100 % ear detection accuracy. However the performance evaluation has been done only on a small database. It has used only 142 images of UND database. In [7], a technique has been proposed where ear is extracted from range image of 3D profile face with the help of a registered 2D profile face image. In this technique, the ear location is detected in 2D profile face image using the AdaBoost based ear detector and corresponding 3D data is extracted from the co-registered 3D profile image. The technique is computationally expensive. It has been tested on a part of UND Collection J2 database which contains 1780 images. It has shown 99.90 % detection rate on a set of selected 830 images of the database.

It should be noted that except the techniques proposed in [1, 2], all other techniques (such as [3, 5, 7]) need a registered 2D profile face image for ear detection in 3D range image. Moreover, these techniques do not offer any viable mechanism to perform ear detection in the presence of scale and pose (rotation) changes. Also, they are not able to detect left and right ears simultaneously and require prior information or specific training for doing so. Further, they are also not completely automatic and cannot be deployed in real applications.

For a 3D ear recognition system, it is very essential to locate automatically and crop the ear from a whole 3D profile face image which may be affected due to scale and pose variations. However, detection of ears from an arbitrary 3D profile face range image is a challenging problem due to fact that ear images can vary in scale and pose under different viewing conditions.

In this chapter, an ear localization technique has been discussed which has attempted to handle these issues by proposing a scale and rotation invariant technique for automatic ear detection in 3D profile face range images. This technique does not require any registered 2D image for ear detection in 3D. Also, it can detect left and right ear at the same time without imposing any additional computational cost.

Rest of the chapter is organized as follows. Section 4.2 discusses the ear localization technique. Scale and rotation invariance of this technique is discussed in Sect. 4.3. Experimental results are analyzed in Sect. 4.4.

4.2 3D Ear Detection Technique

The technique is based on the fact that in a 3D profile face range image, ear is the only region containing maximum depth discontinuities; as a result, this place contains larger edge density as compared to other areas. Further, edges belonging to an ear are curved in nature. The technique consists of three main tasks: preprocessing, ear candidate set computation and ear localization. These steps are similar to that used in ear detection in 2D. Figure 4.1 shows overall flow chart of the technique.

4.2.1 Preprocessing

It consists of four major steps. 3D profile range image is converted to depth map image. Further, edge computation is carried out on the depth map image. These edges are approximated using line segments. Finally, all irrelevant edges are pruned out.

4.2.1.1 Computation of Depth Map Image

The 3D data of profile face used in this study is collected by non-contact 3D digitizer Minolta Vivid 910 which produces 3D scanned data in the form of point cloud grid

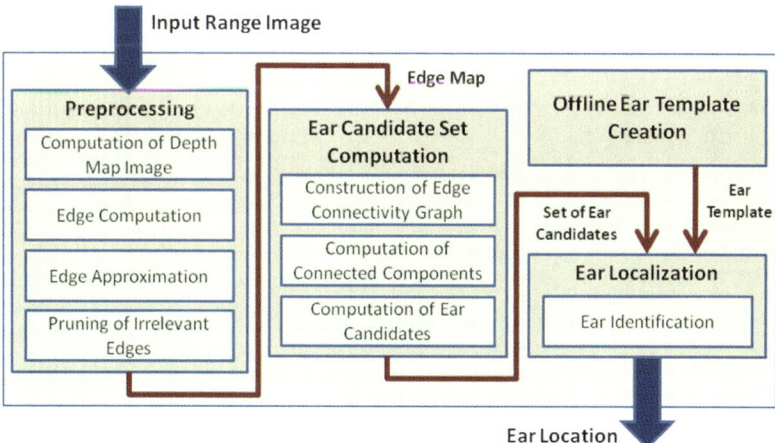

Fig. 4.1 Flow chart of 3D ear detection technique

of size $m \times n$ with each point having a depth information. The digitizer assigns a
large value of depth (*Inf*) if it fails to compute the depth information for a point. Let
$z_{(i,j)}$ be the depth information for a point (i, j) and it contains a real finite value if
depth could be computed; otherwise it is set to *Inf*. In the technique discussed here,
a 3D range image of a profile face is converted to a depth map image which is used
for edge detection. Depth map image $I_{2D} \in R^{m \times n}$ from a range image is obtained
by treating the depth value of a 3D point as its pixel intensity is given as

$$I_{2D}(i, j) = \begin{cases} z_{(i,j)}, & \text{if } z_{(i,j)} \text{ is finite} \\ 0, & \text{otherwise} \end{cases}$$

Pixel intensities in the depth map image are normalized in the range 0–1 using
min-max normalization as defined by

$$I_{2D} = \frac{I_{2D} - min(I_{2D})}{max(I_{2D}) - min(I_{2D})} \qquad (4.1)$$

Steps of computation of depth map image are summarized in Algorithm 4.1.
Figure 4.2a shows few examples of 3D range images where corresponding depth
map images are shown in Fig. 4.2b. It can be noted that the technique to convert a
3D range image to 2D depth image works well for any general pose and rotation of
a 3D profile face. Figure 4.3 shows examples of successful conversion of 3D range
images to depth images for various views of a subject.

Algorithm 4.1 Computation of Depth Map Image

- **Input:** 3D range image $I_{3D} \in R^{m \times n \times 3}$ where $I_{3D}(i, j, :)$ is a 3 elements vector $[x(i, j), y(i, j), z(i, j)]$ which states the 3D information of (i, j)th location in the point cloud grid.
- **Output:** Depth map image $I_{2D} \in R^{m \times n}$.

1: Get the depth information of all the grid points, *i.e.* $z = I_{3D}(:, :, 3)$
2: **for** $i = 1$ to m **do**
3:   **for** $j = 1$ to n **do**
4:     **if** $z_{(i,j)}$ is finite **then**
5:       $I_{2D}(i, j) = z_{(i,j)}$
6:     **else**
7:       $I_{2D}(i, j) = 0$
8:     **end if**
9:   **end for**
10: **end for**
11: /*Normalization of the intensities in the range 0 and 1 */
12: Compute $I_{2D} = \dfrac{I_{2D} - min(I_{2D})}{max(I_{2D}) - min(I_{2D})}$
13: Return I_{2D}.

(a)

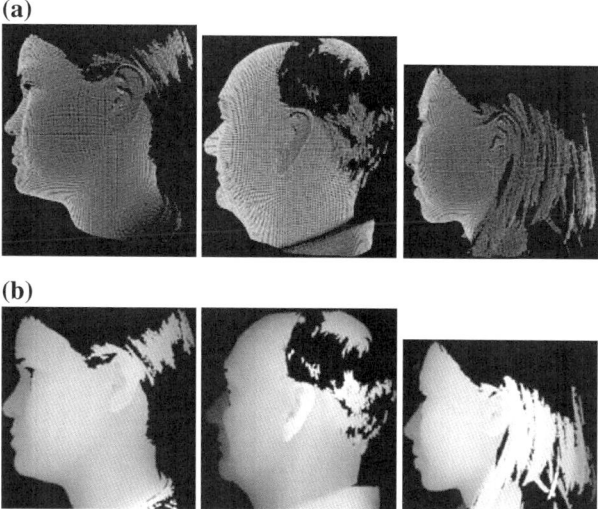

(b)

Fig. 4.2 Examples of 3D range and corresponding depth map images. **a** 3D range images. **b** Depth map images

(a)

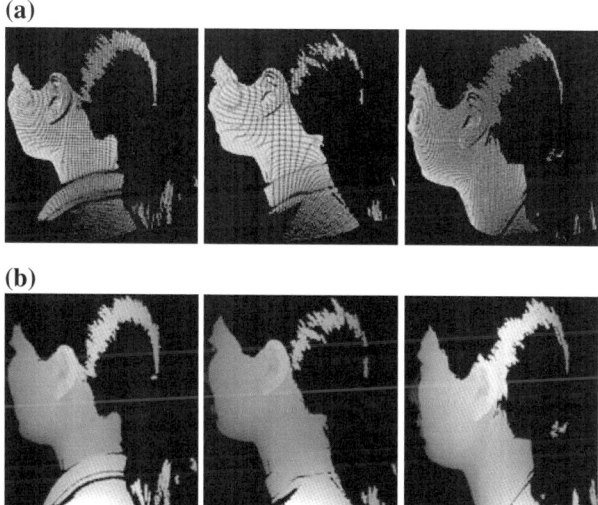

(b)

Fig. 4.3 Original 3D range images and corresponding depth map images for various poses of profile face of a subject. **a** Original 3D range images. **b** Depth map images

4.2.1.2 Edge Computation

Computation of edges in an intensity image is a challenging task. However in a depth map image of a profile face, it is relatively easy due to the presence of strong depth discontinuities in ear region, particularly around the helix of the ear. In the 3D ear detection technique discussed here, edges of depth map image of a profile face are detected using Canny edge operator. Subsequently, a list of all detected edges is obtained by connecting the edge pixels into a list of coordinate pairs. Wherever an edge junction is found, the list is concluded and a separate list is created for each of the new branches. Edges formed due to noise are eliminated using an edge length based criterion where length of an edge is defined as the number of pixels participating in it. Length threshold for edge elimination is chosen automatically proportional to the width of the profile face depth map image. Formally, for an image of width n, the threshold τ can be defined by as $\tau = \kappa n$ where κ is a constant whose value is chosen experimentally. It is chosen to be 0.03 in our experiments.

4.2.1.3 Edge Approximation

All pixels of a computed edge may not be equally important and may not necessarily represent the edge. Inclusion of such pixels in further computation not only may create redundancy but may also slow down the speed of computation. Hence to speedup the processing, redundant pixels are removed by fitting line segments to the edges. In each array of edge points, value and position of the maximum deviation from the line that joins the endpoints is computed. If the maximum deviation at a point is found to be more than the allowable tolerance, the edge is shortened to that point and procedure is repeated to fit the line in the remaining points of the edge. This breaks each edge into line segments where each segment represents the original edge with the specified tolerance.

4.2.1.4 Pruning of Irrelevant Edges

All linear edges (i.e. which need only two points for their representation after line segment fitting) can be removed. As stated earlier in case of 2D ear detection, this is done due to the fact that human ear edges contain some curvature and need at least three points for their representation. Let the set of all edges which may belong to the ear be S.

4.2.2 Ear Candidate Set Computation

It is used to compute the connected components in the graph to obtain ear candidate set. The process of computing such a candidate set consists of three major steps; it builds an edge connectivity graph, obtains connected components in the graph and computes ear candidate set.

4.2.2.1 Construction of Edge Connectivity Graph

Let the set S contain n edges which define the edge map of the profile face depth map image. Let the ith edge e_i in S be defined by a point $p_i \in P$. Let there be a convex hull $CH(e_i)$ defined for each edge e_i. Let there be a newly defined edge connectivity graph $G = (V, E)$ where the set of vertices, V and the set of arcs,[1] E, can be defined by $V = \{p_i \mid p_i \in P\}$ and $E = \{(p_i, p_j) \mid CH(e_i) \text{ intersects } CH(e_j)\}$ respectively.

Due to convexity of edges and the nature of outer ear edges containing inner ear edges, convex hulls of outer edges cut across those of inner edges. Experimentally, it is observed that the convex hull of an ear edge cuts at least one other convex hull of the edge belonging to the ear. Thus, it is expected that all vertices belonging to the ear part in G get connected to one another directly or through another vertex. Since the characteristic of outer edge containing inner edge is not true for the edges belonging to non-ear parts of the profile face depth map image, vertices belonging to non-ear part mostly remain isolated in the graph.

4.2.2.2 Computation of Connected Components

Connected components of graph G are analyzed to localize the ear. Only the connected components with two or more vertices are considered in computing the probable ear candidates. This is due to the fact that any component with single vertex represents single intensity edge in the depth map image and cannot represent an ear; hence it is removed from the graph. Figure 4.4 presents an example of edge connectivity graph and connected components labeling. Figure 4.4a shows an edge image obtained from a profile face depth map image while Fig. 4.4b shows edge connectivity graph constructed for this edge image. Connected components in Fig. 4.4b with more than one vertices are shown inside rectangles.

4.2.2.3 Computation of Ear Candidates

Let $K = \{K_1, K_2, \ldots, K_m\}$ be the set of connected components of graph G where each component has two or more number of vertices. Average vertex degree of a connected component K_j is defined as:

$$d(K_j) = \frac{\Sigma_{i=1}^{n_j} d(p_i)}{n_j}$$

where $d(p_i)$ is the degree of vertex p_i and n_j is the total number of vertices in component K_j. To further discard the false connected components, only components

---

[1] As stated earlier, in this book, "arc" signifies the notion of an edge in a graph. The word "edge" is used in the context of an edge in an image which is a set of connected pixels representing points of high intensity gradient in the image.

Fig. 4.4 An example of
edge connectivity graph and
connected components
labeling. **a** Edge map
(different colors used to
distinguish edges). **b** Edge
connectivity graph
(connected components with
more than one vertex
inclosed inside rectangle).
c Detected ear

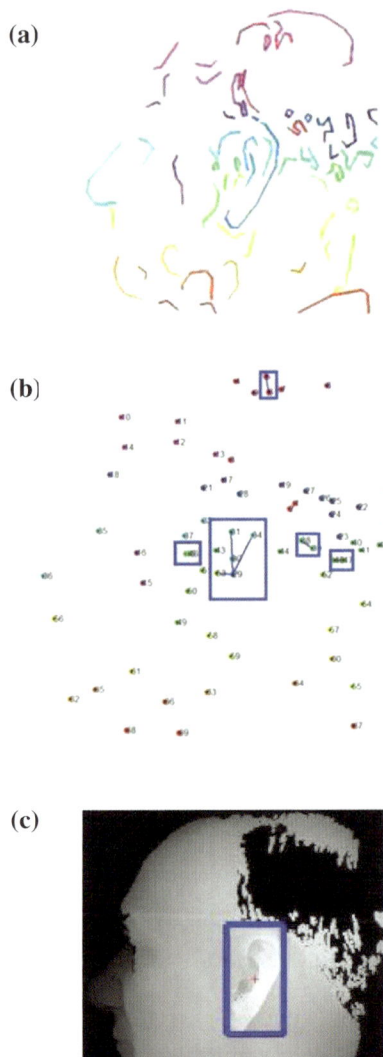

having average vertex degree greater than one are considered to obtain probable ear candidates. This is due to the fact that ear portion of the profile face depth map image is rich in edges due to large depth variation present in this region and hence it is less probable that a connected component representing an ear has only two vertices or average vertex degree one.

A probable ear candidate in a profile face image is defined as the part of the 3D range image cropped using bounding box of the edges participating in a connected component. It is important to note that probable ear candidates are cropped from the 3D point cloud image and not from depth map image. Since the depth map image is obtained from the 3D range image by considering depth value z as intensity, they are registered in x-y plane and hence the bounding box of connected component edges (which are obtained from depth map image) refers to the same portion in depth map image as well as in 3D range image. Ear candidate set is computed by using all connected components of the edge connectivity graph satisfying the criterion of the average vertex degree.

4.2.3 Ear Localization

It is carried out by identifying the true ear among the probable ear candidates with the help of an ear template created off-line. Identification is performed by comparing the ear template with the probable ear candidates.

4.2.3.1 Ear Template Creation

To identify true ear, a template which exhibits the characteristics of scale and rotation invariance is used. To compute such a template in this technique, *shape distribution*, a 3D shape descriptor [8], which is invariant to rotation and scale is used. Shape distribution provides good distinctive features, particularly for ears and non-ears and at the same time it is robust to changes in viewing angle, rotation and scale. It represents shape of a 3D model as a probability distribution sampled from a *shape function* measuring geometric properties of the 3D model. We use L_2-norm as a shape function for computing the shape distribution. It represents the distribution of Euclidean distances between pairs of randomly selected points on the surface of a 3D model. While the distribution describes the overall shape of the object efficiently, samples from this distribution can be obtained quickly and easily. Figure 4.5 shows the shape distribution of 200 ear and 200 non-ear (facial part) samples. It can be clearly envisaged that the shape distribution of ears is condensed in a small area as shown in red color. However, it is not the case with the non-ears.

To compute an ear template, a training set is created by selecting randomly a few 3D ear samples and their shape distributions are obtained. The ear template is obtained by computing the average shape distribution by taking the average of respective bins of all shape distributions representing the training ears.

4.2.3.2 Ear Identification

Let the ear candidate set be $I_E = \{I_1, I_2, \ldots, I_\eta\}$ where η is the cardinality of set I_E and I_k is the portion of the 3D profile face range image representing kth probable ear candidate, $k = 1, 2, \ldots, \eta$. Identification of true ear is performed by comparing the

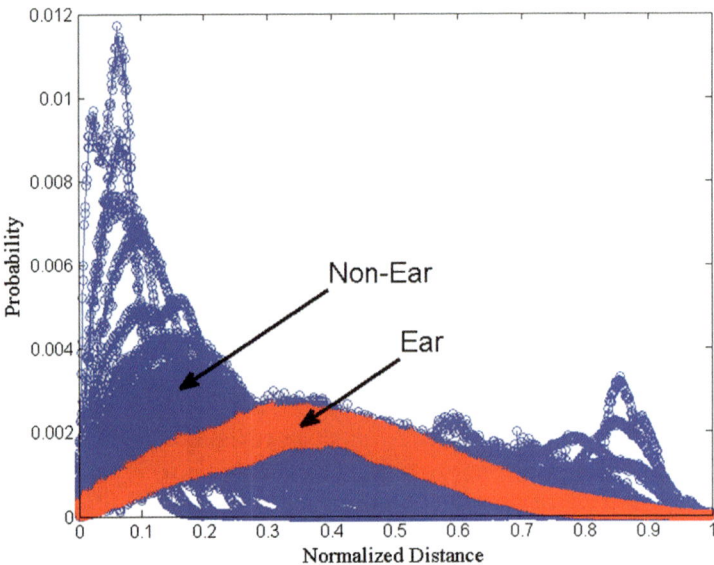

Fig. 4.5 Shape distributions of 200 ear (shown *red*) and 200 non-ear (shown *blue*) samples

ear template with the shape distributions of the ear candidates in I_E. Comparison is performed using Bhattacharyya distance D which is defined as follows for two shape distributions f and g

$$D(f, g) = 1 - \sum_{i=1}^{n} \{\sqrt{f_i g_i}\}$$

where f_i and g_i are the number of samples of f and g in the ith bin and n is the total number of bins in the distributions. Let $T_E = \{T_{I_1}, T_{I_2}, \ldots, T_{I_n}\}$ be the shape distribution set for the ear candidates in I_E. Bhattacharyya distance is computed between ear template (T) and all elements in T_E to obtain a match score vector *MatchScore*. The true ear candidate I_ξ is obtained such that

$$\xi = \arg \min_{i} \{MatchScore[i]\} \tag{4.2}$$

That means, the ear candidate from I_E for which Bhattacharyya distance is minimum, is declared as the true ear candidate. Figure 4.4c shows the true ear location enclosed by a rectangle obtained after ear identification.

4.3 Scale and Rotation Invariance

A very important and useful characteristics of the 3D ear localization technique discussed in this chapter is that it is inherently scale and rotation invariant.

4.3.1 Invariance to Scale

Crux of the this technique lies in the manner in which the edge connectivity graph G is constructed. Construction of a good graph depends on the criterion used to define the connectivity among the vertices. To make the ear detection scale invariant, criterion to connect vertices in G should also be scale invariant. In the technique, it is based on the intersection of convex hulls which can define connectivity among the vertices irrespective of the scale of the ear image. This makes the technique scale invariant.

To show the scale invariance, a synthetic edge map as shown in Fig. 4.6a is considered. It is scaled-down as shown in Fig. 4.7a. Convex hulls and edge connectivity

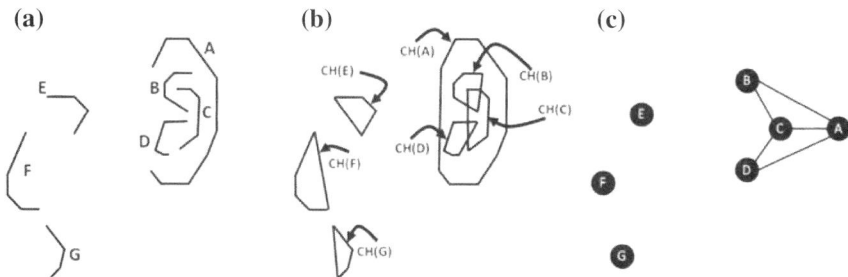

Fig. 4.6 An example of construction of edge connectivity graph for an edge map. **a** Edge map. **b** Convex hulls. **c** Edge connectivity Graph

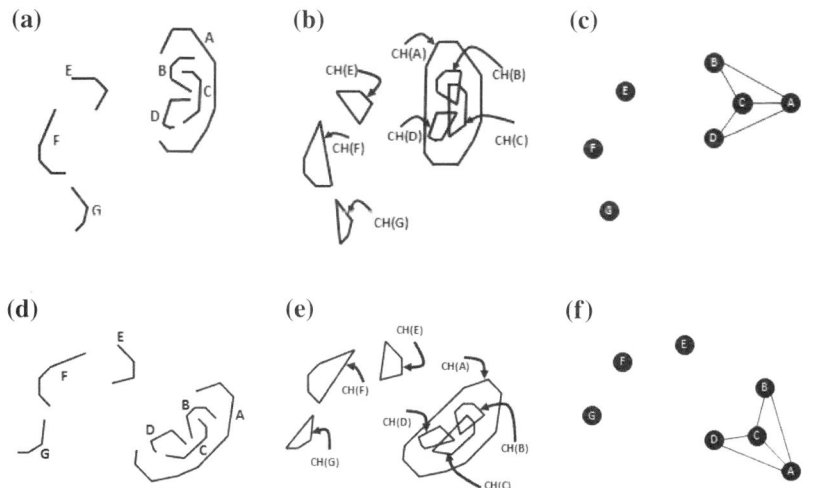

Fig. 4.7 Demonstration of scale and rotation invariance: **a** scaled-down edge map of Fig. 4.6a, **b** convex hulls of the edges shown in (**a**), **c** edge connectivity graph for (**b**), **d** rotated edge map of Fig. 4.6a, **e** convex hulls of the edges shown in (**d**), **f** edge connectivity graph for (**e**)

graph for this edge map are depicted in Fig. 4.7b, c respectively. It is apparent that the obtained graph is similar to one shown in Fig. 4.6c. Similar results have been observed for the scaled-up version of the edge map.

4.3.2 Invariance to Rotation

Since structural appearance of the ear (hence the edge map) does not change due to in-plane rotation, criterion based on convex hull can be used to define connectivity among the vertices in the graph G, even in the presence of rotation. This makes the technique rotation invariant. The technique can also detect ear in the presence of out-of-plane rotation. However, it is expected that such type of rotation does not entirely hide the structural details of the ear.

   To show the rotation invariance, synthetic edge map shown in Fig. 4.7a is rotated by $-45°$ as shown in Fig. 4.7d. The convex hulls and edge connectivity graph for the obtained rotated edge map are shown in Fig. 4.7e, f respectively. It can be seen that the obtained graph is similar to one shown in Fig. 4.7c. Similar results can be demonstrated for any amount of rotation.

4.4 Experimental Results

Experiments are conducted on Collection J2 (UND-J2) [5, 9] of University of Notre Dame which contains 3D profile face images. Images where camera could not capture the 3D information of the ear correctly are discarded and 1604 images are used for experiments. These images are influenced by scale and pose variations. Also, some of the images are occluded by hair and ear rings. Figure 4.8 shows few segmentation results on UND-J2 database.

4.4.1 Performance Evaluation

Performance of the ear detection technique discussed in this chapter is shown in Table 4.1. It produces 99.38 % correct detection rate. To show its robustness against rotation, there are two new test data sets generated by performing in-plane rotation of $+90°$ and $-90°$ on the original 3D range images. Performance of ear detection on both data sets is found to be same as the one obtained on the original set. Also, to show that the technique can detect left and right ears at the same time, another test is conducted by flipping the 3D profile face range images horizontally, thus making all left profile face images right. We have achieved the same performance on this experiment too. Further, to test the technique against scale variations, 194 3D profile

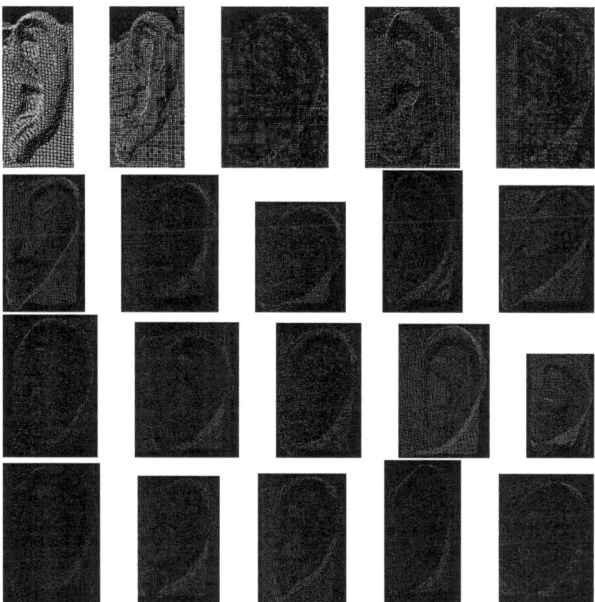

Fig. 4.8 Few results of 3D ear detection from UND-J2 database

Table 4.1 Ear detection accuracy on UND-J2 database

Test images	Accuracy (%)	Remarks
1604	99.38	Normal 3D range images
1604	99.38	Images rotated by +90°
1604	99.38	Images rotated by −90°
1604	99.38	Images flipped horizontally
194	100	Images of varied scales
149	99.32	Images without of plane variations

face range images of different scales are picked out from UND-J2 database. It has been been found that the technique achieved 100 % detection rate in this experiment.

A set of 149 range images where images are affected by out-of-plane rotation is formed from the UND-J2 database to test the technique against out-of-plane rotations. Detection accuracy in this case is found to be 99.32 %. The technique has failed to detect ear only in one range image due to acute rotation which made the structural details of the ear invisible. Ear detection accuracies for the test cases, shown in Table 4.1, have been achieved without altering any parameter or performing any specific training. It can be noted that scanning error such as small holes created due to occlusion during the scanning process does not affect the performance. However, it may deteriorate if significant portion of the outer helix of the ear is occluded due to hair.

4.4.2 Comparison with Other Techniques

Performance of the presented technique has been compared with some of the well known techniques. The technique proposed in [5] has achieved detection accuracy of 85.54 % on a part of UND Collection J2 database when only 3D range information is used. It has achieved 100 % ear detection when information from both 2D and 3D images is used. However, it is important to note that the test set used in this study is very small and consists of a few selected images of UND-J2 database. Its performance on whole UND-J2 database is not reported. Moreover, this technique is not rotation invariant as its performance very much dependent on the correctness of the nose tip detection. In case of rotation, heuristics used to locate nose tip may result into a wrong ear detection. In [3], UND databases Collection G and a part of Collection F (which are the subsets of UND Collection J2) have been used. Ear detection accuracy is reported to be 87.71 % on the test data of 700 images. The technique has used 3D range images along with 2D to perform ear detection. When only 3D information is used, its performance is not reported in [3]. However, it is apparent that the performance of the technique using only 3D information cannot be superior to that obtained using 2D and 3D information both. Moreover, the technique is based on template matching and may suffer in presence of rotation and scale variations.

The technique proposed in [7] claims to achieve good 3D ear detection accuracy. However, it is not a 3D ear detection technique in true sense because it does not use the 3D information of the ear for detection. Instead, it detects ear on a registered 2D image and uses the location of the ear in 2D image to locate ear in 3D. Hence, it cannot be employed when only 3D range data of the profile face is available. Also, if the test images are rotated or their appearance changes with respect to training data, this technique may fail because training images may not include such cases. Creating a database of ears with all possible rotations may require a very large storage space as well as huge amount of training time. Moreover, to detect ear of different scales, there is a need to perform an exhaustive search with filters of various size which is computationally expensive. Besides this, though UND-J2 database contains 1780 images, this technique separates a part of it (830 images) to show the performance rather than using the whole UND-J2 database. Size of the test data used in all of these techniques [3, 5, 7] is very small as compared to the test data used in our experiments.

On the other side, the technique discussed in this chapter is inherently capable of handling rotation (pose) and scale changes and does not demand any extra compu- tational cost to accomplish this. Also, it can detect left and right ears simultaneously without any prior information of the image or specific training. It does not need a registered 2D image to detect ear in 3D. Independence from 2D image makes the technique generic and applicable to real life scenarios. Experimental results of the

Table 4.2 Performance comparison on UND-J2 database

Technique	Database size	Images used	Accuracy (%)	Remarks
[5]	1780	415[a]	85.54	Only 3D range images used
			100	3D range images used with 2D
[3]	700	700	87.71	3D range images used with 2D[b]
[7]	1780	830[c]	99.90	Use of 2D registered images
Technique discussed	**1780**	**1604[d]**	**99.38**	**Only 3D range images used**

[a]Test images selected manually
[b]No results reported in [3] when only 3D range images are used
[c]Test images selected manually
[d]All images of UND-J2 except the one where 3D ear information could not be captured properly

technique which has been tested on a larger data set are more stable as compared to the one reported in [3, 5, 7].

From Table 4.2, it can be observed that the technique discussed in this chapter performs better than the techniques in [3, 5] (when it uses only 3D information). Also, it performs competitively with respect to the techniques proposed in in [5, 7] (when it uses 2D images along with 3D) but it is important to note that these techniques have been tested only on 415 and 830 images respectively. In [7], detection is carried out using 2D registered image and not actually using the 3D data. Our technique can achieve comparable ear detection performance for 3D range images by using only 3D range information. A comparison of our technique with the most recent technique in [7] is given in Table 4.3.

Table 4.3 Comparison with [7]

Parameters	Techniques	
	Islam et al [7]	Technique discussed in this chapter
Training overhead	More, training with 1000 s of positive and negative samples	Only for ear template creation
Training time required	Few days	Few minutes
Inherently invariant to rotation and scale	No	Yes
Registered 2D image required	Yes	No
Can work directly on 3D data	No	Yes
Can detect left and right ear without prior knowledge	No	Yes

References

1. Chen, Hui and Bir Bhanu. 2004. Human ear detection from side face range images. In *Proceedings of International Conference on Pattern Recognition (ICPR'04)*, vol. 3, 574–577
2. Chen, Hui, Bhanu, Bir 2005. Shape model based 3D ear detection from side face range images. In *Proceedings of International Conference on Computer Vision and Pattern Recognition-Workshop*, 122–127
3. Chen, Hui, and Bir Bhanu. 2007. Human ear recognition in 3D. *IEEE Transactions on Pattern Analysis and Machine Intelligence* 29(4): 718–737.
4. Yan, Ping and Kevin W. Bowyer. 2005. Empirical evaluation of advanced ear biometrics. In *Proceedings of International Conference on Computer Vision and Pattern Recognition-Workshop*, vol. 3, 41–48
5. Yan, Ping, and K.W. Bowyer. 2007. Biometric recognition using 3D ear shape. *IEEE Transactions on Pattern Analysis and Machine Intelligence* 29(8): 1297–1308.
6. Zhou, Jindan, Cadavid, S., and Abdel-Mottaleb, M. 2010. Histograms of categorized shapes for 3D ear detection. In *Proceedings of IEEE International Conference on Biometrics: Theory Applications and Systems (BTAS' 10)*, 1–6
7. Islam, S.M.S., Rowan Davies, Mohammed Bennamoun, and Ajmal S. Mian. 2011. Efficient detection and recognition of 3D ears. *International Journal of Computer Vision* 95(1): 52–73.
8. Osada, Robert, Thomas Funkhouser, Bernard Chazelle, and David Dobkin. 2002. Shape distributions. *ACM Transactions on Graphics* 21(4): 807–832.
9. University of Notre Dame Profile Face Database, Collection J2. http://www.nd.edu/~cvrl/CVRL/DataSets.html

Chapter 5
Ear Recognition in 3D

5.1 Introduction

Three-dimensional (3D) images are found to be useful to design an efficient biometric system because they offer resilience to problems such as sensitivity towards pose, illumination and scale variations, common in two-dimensional (2D) images. Further, cost of 3D scanners has been drastically reduced. As a result, efforts are being made to design efficient recognition systems based on 3D images.

In spite of ear having numerous rich characteristics as compared to other rival biometric traits, poor accuracy of a 2D ear recognition system has kept it away from practical applicability. However, the use of 3D ear images has helped in enhancing the recognition accuracy. Recently there exist several ear recognition techniques which are based on either only 3D ear data or both 3D and 2D ear data. In [1], a two-step ICP algorithm has been proposed to match between two 3D ears. The ICP algorithm is used to find the initial rigid transformation to align helix of one 3D ear image with that of the other image. This initial transformation is applied to the selected locations of the ears and ICP algorithm is further used to refine iteratively the transformation to bring ears into best alignment. This technique has achieved rank-1 recognition rate of 93.3 % with 6.7 % of *EER* in a small 3D ear database consisting of 30 subjects. In [2], an ear based human recognition technique which has made use of 2D and 3D ear data has been studied. The technique has explored the use of PCA (Eigen-ear) approach with 2D and 3D ear images, Hausdorff matching of edge images obtained from 3D images and ICP matching of the 3D data. It has been concluded that ICP based matching achieves the best performance and shows good scalability with the size of the database.

Another technique proposed in [3] has also made use of 2D and 3D ear data for recognition. It has shown improvement using multi-algorithmic based system over the unimodal system. Further, it has proposed a fusion rule using the interval distribution between top two ranks. It has been observed that multimodal 2D PCA together with 3D ICP has shown the best performance and has achieved rank-1 recognition rate of 91.7 % in the database of 942 pairs of $2D$ and $3D$ images collected from 302 subjects.

© Springer Science+Business Media Singapore 2015
S. Prakash and P. Gupta, *Ear Biometrics in 2D and 3D*,
Augmented Vision and Reality 10, DOI 10.1007/978-981-287-375-0_5

In [4], local surface descriptors of two ear images have been compared to obtain a correspondence between two local surface patches. The rank-1 recognition rate of this technique has been reported as 90.4 % on a 3D ear database of 104 images collected from 52 subjects. All these techniques have not only been evaluated on small and comparatively less challenging databases but also have poor recognition rates. Moreover, mostly their performance fall when tested on larger databases as reported in [2] where 3D ICP performance falls from 92 to 84.1 % when database size is changed from 25 to 302 subjects.

In [5], an ear based system for human recognition has been proposed which includes automated segmentation of the ear in a profile view image and ICP based 3D shape matching for recognition. The technique has achieved 97.8 % rank-1 recognition rate with 1.2 % *EER* on UND-J2 database. Since it has made use of nose tip and ear pit boundary for ear detection, it may not work properly if the nose tip or the ear pit is not clearly visible. However, due to pose variations, one cannot always assume the visibility of nose tip or ear pit. Another ear based recognition technique which includes automatic ear segmentation has been proposed in [6]. Ear detection in this technique has been carried out using a single reference 3D ear shape model. The technique has proposed two representations of the ear for recognition. These include the ear helix/antihelix representation obtained from the detection algorithm and the local surface patch (LSP) representation computed at salient feature points. ICP algorithm is employed for final probe and gallery image matching. The technique has obtained 96.8 % rank-1 recognition rate on Collection F of the UND database (UND-F). It has also been tested on University of California Riverside (UCR) database-ES1 which is comprised of 310 frontal ear images of 155 subjects collected on the same day. On this database, it has achieved rank-1 recognition rate of 94.4 %. It has assumed perfect ear detection. Otherwise, manual segmentation of the ear contour is performed prior to recognition. An ear recognition system proposed in [7] has used a generic annotated ear model to register and to fit each ear data. It has extracted a compact biometric signature of the ear that retains 3D information for recognition. It has used ICP and Simulated Annealing [8] algorithms to register and to fit probe and gallery ear data. It has achieved 93.9 % rank-1 recognition rate on UND-J2 database. However, this technique takes large amount of time for enrollment and few minutes for matching. In [9], 3D ear recognition which has used structure from motion and shape from shading techniques has been proposed. Ear segmentation has been done with the help of interpolation of ridges and ravines identified in each frame in a video sequence. Rank-1 recognition rate has been obtained as 84 % on a small database consisting of 61 gallery and 25 probe images.

In [10], Adaboost has been used for ear segmentation. It has considered 2D and 3D ear data. Matching of the ears has been performed on a coarse-to-fine hierarchical alignment through ICP algorithm. The technique has achieved 93 % rank-1 recognition rate on a database of 200 pairs of $2D$ and $3D$ ear images obtained from 100 subjects. In [11], there exists another ear based recognition technique which is based on local 3D features computed at distinctive locations of the 3D ear data by considering an approximated neighborhood surface around these locations. It has established the correspondences between features and has used these correspondences to align two

data sets. Finally, ICP algorithm is used for final recognition. It has achieved 90 %
rank-1 recognition rate on 200 3D ear range images collected from 100 subjects. The
technique proposed in [12] has first detected ear from 2D profile face image using
the Cascaded AdaBoost detector and then has extracted corresponding 3D ear data
from co-registered 3D image. It has computed local 3D features from the ears and
has used them for initial matching whereas ICP based technique has been used for
final matching. It has achieved 93.5 % rank-1 recognition rate with *EER* of 4.1 % on
UND-J2 database. The techniques proposed in [10–12] have mostly performed low
in case of occlusions and large pose variations.

Collaboration of face and ear is a good choice of biometric traits because of
their physiological structure and location. Also, both of them can be acquired
non-intrusively. To exploit these advantages, there exist few multi-biometric techniques which are based on ear and face. In [13], a unified technique that fuses 3D
ear and facial data has been proposed. It has used an annotated deformable model
of the ear and the face to extract respective geometry images from the data. It has
computed wavelet coefficients from the geometry images and has used these coefficients as biometric signature for recognition. The technique has achieved 99.7 %
rank-1 recognition rate on a multimodal database of 648 pair of 3D face and ear
range images obtained from 324 subjects. In [14], a technique based on local 3D
features to fuse ear and face biometrics at score level has been proposed. It has been
tested on a multimodal database which includes 326 gallery images with 311 and 315
probes images with neutral and non-neutral expressions respectively. It has achieved
a recognition rate of 98.71 % and a verification rate of 99.68 % for fusion of the
ear with neutral face. Further, a recognition rate of 98.10 % and a verification rate
of 96.83 % has been achieved when facial images have expressions. There exists
another technique in [15] which has used 3D morphable model of the head and the
ear for human recognition. It has been evaluated on 160 training samples to compute
its performance. Though ear and face based multimodal techniques have achieved
improved performance, they are computationally expensive for the large volume of
3D ear and face data and hence have low practical applicability.

A general 3D object recognition technique by combining local and holistic features has been proposed in [16] and has been evaluated for 3D ear recognition task.
It has primarily focused on local and holistic feature extraction and matching components, in addition to fusion framework to combine these features at the matching
score level. It has yielded a rank-1 recognition rate of 98.6 % and an *EER* of 1.6 %
on UND-J2 database.

Most of these techniques make use of ICP algorithm to match the 3D ear images
by aligning one ear with other and to consider alignment error as the match score.
Since performance of ICP based matching highly depends on the initial state of
the two data sets to be matched, these recognition techniques do not produce good
alignment in most of the cases and hence lead to poor accuracy.

This chapter presents an efficient human recognition technique which uses 3D ear
images along with their respective co-registered 2D ear images. It presents a two-step
matching technique to compare two 3D ear images. It first coarsely aligns the 3D ear
images using few salient 3D data points which are computed with the help of local

2D feature points obtained from co-registered 2D ear images. In the second step, it uses a technique obtained by integration of Generalized Procrustes Analysis (GPA) [17] with ICP (GPA-ICP) for final alignment. Rest of the chapter is organized as follows. Section 5.2 discusses Generalized Procrustes Analysis which has been used in the 3D ear recognition technique discussed in this chapter. Next section presents the ear recognition technique. Experimental results are analyzed in Sect. 5.4.

5.2 Generalized Procrustes Analysis

Procrustes Analysis is used to align two sets of data points. Let A and B be two matrices of data points, each of size $p \times q$ where p is the number of data points with each one represented in q dimensional space. The Orthogonal Procrustes Analysis [18] computes an orthogonal transformation matrix T which minimizes the residual matrix $E = AT - B$. This technique has been extended in [19] to compare two sets of data points in presence of rotation, scale and translation by introducing a rotation matrix R, a translation vector t and a scaling factor s. This generic method is commonly known as Extended Orthogonal Procrustes Analysis (EOPA). Residual matrix E minimized in this case is defined as $E = sAR + u \otimes t - B$ where u is a $p \times 1$ column vector of all ones, t is a $1 \times q$ row vector for translation and $\otimes$ denotes Kronecker product.

Weighted version of EOPA, known as Weighted Extended Orthogonal Procrustes Analysis (WEOPA), can be obtained by assigning weights to points and their components. Resulting residual to be minimized in this technique is given by

$$e = tr \left[(sAR + u \otimes t - B)^T W^P (sAR + u \otimes t - B) W^Q \right] \qquad (5.1)$$

where W^P and W^Q are the diagonal weight matrices of size $p \times p$ and $q \times q$ respectively. Matrix W^P contains the weights for data points whereas W^Q consists of the weights for components of data points. Generalized Procrustes Analysis (GPA) [17] is a popular technique which is used to align more than two sets of data points. Residual objective function to be minimized to get the solution for GPA is given by

$$e = tr \left[\sum_{i=1}^{n} \sum_{j=i+1}^{n} ((s_i X_i R_i + u \otimes t_i) - (s_j X_j R_j + u \otimes t_j))^T \right.$$
$$\left. ((s_i X_i R_i + u \otimes t_i) - (s_j X_j R_j + u \otimes t_j)) \right] \qquad (5.2)$$

where $X_1, X_2, \ldots, X_n$ are n sets of data points, each having p points in a q dimensional space. Let $X_i^P = s_i X_i R_i + u \otimes t_i$. Then GPA Eq. 5.2 can alternatively be given by

$$\sum_{i<j}^{n} \|X_i^p - X_j^p\|^2 = \sum_{i<j}^{n} tr\left[(X_i^p - X_j^p)^T (X_i^p - X_j^p)\right] \qquad (5.3)$$

It has been shown in [20] that Eq. 5.3 can be equivalently written as

$$n\sum_{i<j}^{n} \|X_i^p - C\|^2 = n\sum_{i<j}^{n} tr\left[(X_i^p - C)^T (X_i^p - C)\right] \qquad (5.4)$$

where C is called geometrical centroid and is estimated as $C = \frac{1}{n}\sum_{i=1}^{n} X_i^p$.

Equation 5.4 can be solved iteratively to get the transformation parameters s_i, R_i and t_i, $i = 1, \ldots, n$. To get the solution, first centroid C is initialized. At every intermediate step, transformation parameters for each set of data points are directly computed with respect to C using WEOPA for $W^Q = I$. After getting the transformation parameters, all sets of data points are updated and obtained values are used to compute new centroid. This process is repeated until global convergence for centroid C is obtained. To handle a practical situation where all sets of data points may not have same number of points, a binary diagonal matrix M_i of size $p \times p$ is used [21]. The value of diagonal element in this matrix is set to 1 for existence and 0 for absence of a data point in the ith set. Modified GPA Eq. 5.4 for this case can be given by

$$n\sum_{i<j}^{n} \|X_i^p - C\|^2 = n\sum_{i<j}^{n} tr\left[(X_i^p - C)^T M_i (X_i^p - C)\right] \qquad (5.5)$$

where, centroid C is defined as $C = \left(\sum_{i=1}^{n} M_i\right)^{-1} \left[\sum_{i=1}^{n} M_i (s_i X_i R_i + u \otimes t_i)\right]$.

If order to assign weights to data points of ith set, weight matrix W_i^P can be combined with M_i, i.e. M_i can be replaced by $W_i^P M_i$ in Eq. 5.5.

5.2.1 GPA-ICP: Integration of GPA with ICP

GPA and GPA-ICP [22] differ in the way they define the point correspondences between the sets of 3D data points which are matched. GPA needs point correspondences to be defined manually (as done in [23]). However, GPA-ICP proposes to automate the process of defining the point correspondences. Like ICP, it makes use of nearest neighbor points for defining the point correspondences. To get robust and true point correspondences, it considers mutual nearest neighbor of each data point, instead of considering simple nearest neighbor. A data point $x \in A$ is said to be mutual nearest neighbor to data point $y \in B$ if the closest neighbor of x in B is y and the closest neighbor of y in A is x. A set of mutual nearest neighbor data points is called an independent set and is represented by a unique point in C, computed as the centroid of the points of the independent set.

Since in case of biometric recognition, it is required to align only two sets of data points, one for each gallery and probe ear images, the value of n is kept as 2 in GPA-ICP. Also, we have assigned equal weights to all the data points considering $W_i^P = 1, i = 1, \ldots, n$.

5.3 3D Ear Recognition Technique Based on Local 2D Features and GPA-ICP (LFGPA-ICP)

This section presents a technique, named as LFGPA-ICP, for matching of two 3D ear images. It makes use of local 2D feature points and GPA-ICP for matching. It consists of two major steps. In the first step, local 2D feature points obtained from co-registered 2D ear images are used to coarsely aligning 3D ear images. In the second step, GPA-ICP based matching is used to achieve final alignment.

5.3.1 Local 2D Feature Points Based Coarse Alignment

Two 3D ear images which are being matched are first coarsely aligned with the help of local 2D feature points obtained from co-registered 2D ear images. This helps in getting fast and accurate convergence of GPA-ICP in final matching. Moreover, it also helps to avoid GPA-ICP technique from getting stuck into a local minima.

5.3.1.1 Local 2D Feature Points Computation

As we have seen in Sect. 2.2.2, Speeded up Robust Feature (SURF) [24] has been found to be robust and distinctive in representing local image information. It finds some unique feature points along with their respective feature vectors from a given image. A feature point in SURF is described by using the intensity content within the neighborhood of the feature point and by representing it with the help of sum of approximated 2D Haar wavelet components.

Since in LFGPA-ICP (ear recognition technique presented here), the interest lies only on the distinct feature point locations and not on their descriptor vectors, we use only feature point computation part of the SURF to get distinct feature locations. These locations are used in the process to get coarse alignment of the 3D ear images. An example of detected SURF feature points for an ear image is shown in Fig. 5.1.

Fig. 5.1   An example of
SURF feature points
(reproduced from Fig. 2.2)

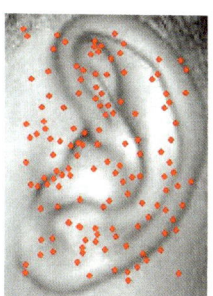

5.3.1.2 Coarse Alignment of 3D Ear Images

Since LFGPA-ICP uses a pair of 2D and 3D ear images co-registered with each other with respect to x-y plane, coordinate locations of the feature points obtained from a 2D ear image in the previous step can be used to fetch corresponding 3D data points from the co-registered 3D ear image. To align the data of two 3D ear images coarsely, this characteristics is used and two sets of salient 3D data points are obtained from 3D ear images with the help 2D feature point locations of their respective co-registered 2D ear images. ICP matching is performed between these two sets of salient 3D data points and a transformation matrix T which aligns them with minimum registration error is obtained. Figure 5.2 shows the steps involved in computation of transformation matrix T. These steps are also summarized in Algorithm 5.1. The transformation matrix T is applied on whole 3D data points of the ear images which are matched to align them coarsely with each other. Steps involved in the alignment are given in Algorithm 5.2.

It is important to note that such an initial coarse alignment is only possible when a 3D ear image is provided along with its co-registered 2D ear image. The reason behind computing salient points of a 3D ear image with the help of local 2D feature points rather than directly computing them from the 3D ear image is that the computation of feature points in 2D is computationally efficient and provides robust feature locations as compared to 3D. Moreover, the field of computation of robust local feature points in 3D is not yet matured.

There exist attempts to utilize the usefulness of local feature descriptors in other biometric applications as well. For example, Bustard and Nixon [25] have used SIFT (Scale Invariant Feature Transform) [26] feature points in 2D ear recognition for the registration of probe and gallery images before matching. In [27], SURF features have been used in computation of a rotation and scale invariant ear template.

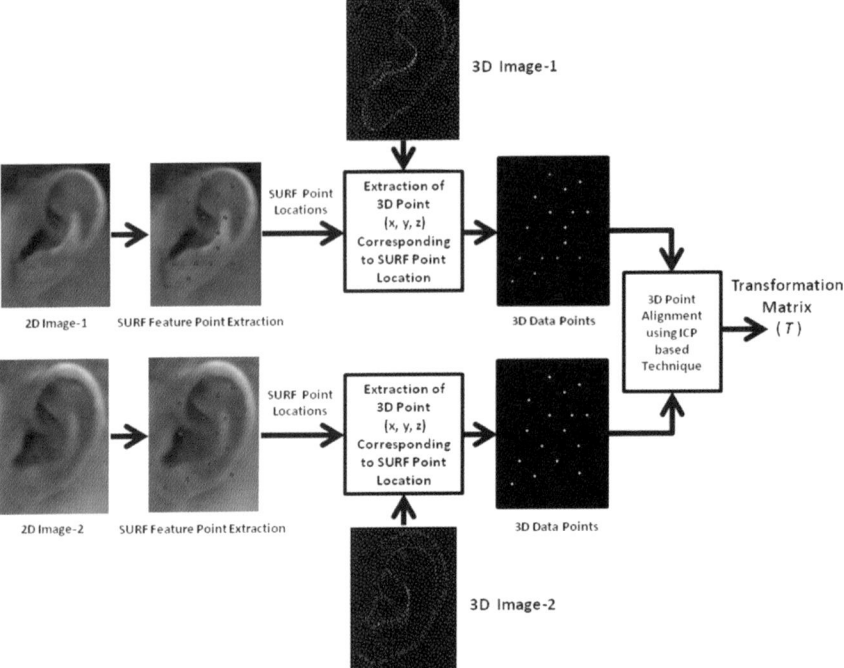

Fig. 5.2 Computation of transformation matrix used in coarse alignment of 3D data

5.3.2 Final Alignment Using GPA-ICP Technique

After achieving coarse alignment, GPA-ICP [22] is employed for final alignment of the 3D ear images being matched. GPA-ICP technique is originally proposed for the alignment of multiple sets of 3D data points (point clouds). However in LFGPA-ICP, its use is restricted to align only two sets of 3D data points as it is required to align/compare only two sets of 3D data points, one each for gallery and probe ears in our case. Steps involved in the final alignment of 3D ear images are summarized in Algorithm 5.3.

The reason behind using the GPA-ICP in place of traditional ICP for final alignment is that it is found to be robust as compared to ICP. This is due to the following reason. In GPA-ICP, 3D data point sets which are aligned are partitioned into independent sets where each independent set contains mutual nearest neighbor points. Further, centroid of the points participating in each independent set is computed and is used in aligning the data points of the independent set. In ICP, alignment of two

---

Algorithm 5.1 Computation of Transformation Matrix

- **Input:** Two 3D ear images $I_{3D}^1 \in R^{m_1 \times n_1 \times 3}$ and $I_{3D}^2 \in R^{m_2 \times n_2 \times 3}$ with their corresponding co-registered 2D ear images $I_{2D}^1 \in R^{m_1 \times n_1}$ and $I_{2D}^2 \in R^{m_2 \times n_2}$. Note that $I_{2D}(i, j)$ states the gray value at pixel (i, j) in image I_{2D} whereas $I_{3D}(i, j, :)$ represents (x, y, z) value at location (i, j) in 3D image I_{3D}.
- **Output:** Transformation matrix T.

1: /*Compute feature point locations.*/
2: **for** $i = 1$ to 2 **do**
3:    Compute $P_i = surfpoints(I_{2D}^i)$ where $surfpoints(.)$ is a function which takes a 2D image and returns a matrix P_i of feature point (key-point) locations of size $l_i \times 2$ where l_i is the number of feature points and a row in P_i represents (x, y) coordinates of a feature point.
4: **end for**
5: /*Extract (x,y,z) points from 3D ear images corresponding to feature point locations of co-registered 2D ear images.*/
6: **for** $j = 1$ to 2 **do**
7:    **for** $k = 1$ to l_j **do**
8:       $X_j(k, :) = I_{3D}^j(P_j(k, 1), P_j(k, 2), :)$
9:    **end for**
10: **end for**
11: /*Computation of Transformation Matrix T*/
12: $T = icp(X_1, X_2)$ where $icp(.)$ is a function which takes two 3D data point matrices of the ears and returns a transformation matrix T which aligns them the best.
13: Return T.

Algorithm 5.2 Coarse Alignment of the 3D Ear Data

- **Input:** Transformation matrix T and two 3D ear images $I_{3D}^1 \in R^{m_1 \times n_1 \times 3}$ and $I_{3D}^2 \in R^{m_2 \times n_2 \times 3}$ which need to be aligned with each other.
- **Output:** Coarsely aligned 3D data point matrices Y_1 and Y_2 corresponding to 3D ear images I_{3D}^1 and I_{3D}^2 respectively.

1: /*3D Ear Image Data*/
2: **for** $k = 1$ to 2 **do**
3:    Compute $Y_k = reshape(I_{3D}^k)$ where $reshape(.)$ is a function which takes 3D matrix of size $m_k \times n_k \times 3$ and reshapes it into a 2D matrix of size $m_k n_k \times 3$ such that $I_{3D}^i(i, j, :)$ (i.e. (x,y,z) value of (i, j)th location in I_{3D}^i) becomes a row in Y_k.
4: **end for**
5: /*Perform transformation of Y_2 to make it coarsely aligned with Y_1*/
6: Compute $Y_2 = Y_2 \times T$
7: Return Y_1 and Y_2.

---

sets of data points is carried out by directly aligning the points of the two sets with each other. It has observed that the alignment of two sets of data points carried out with respect to centroid points provides better alignment as compared to the one achieved by direct alignment of the points of the two data sets with each other.

Algorithm 5.3 Final Alignment of the 3D Ear Data

- **Input:** Coarsely aligned two 3D ear data point matrices Y_1 and Y_2 corresponding to 3D ear images I_{3D}^1 and I_{3D}^2 respectively.
- **Output:** Alignment (registration) error e.

1: /*Transform Y_2 to make it aligned with Y_1 using GPA-ICP */
2: Compute $e = gpa\_icp(Y_1, Y_2)$ where $gpa\_icp(.,.)$ is a function which takes two 3D data point matrices and returns the alignment error after performing the best possible alignment between them using GPA-ICP technique.
3: Return e.

5.4 Experimental Results

LFGPA-ICP has been evaluated on University of Notre Dame public database-Collection J2 (UND-J2) [5]. Images of the database are collected using Minolta Vivid 910 scanner and are affected by scale and pose variations. They have been acquired in two sessions and time gap between the two sessions has been kept at least 17 weeks. It has been observed that there are many images in the database which are affected by occlusion due to hair and ear rings.

The database consists of 2414 3D (with co-registered 2D) profile face images. It has been found that there are many duplicate images in the database. The LFGPA-ICP has been tested on only 1780 3D (and co-registered 2D) profile face images collected from 404 subjects (2 or more samples per subject) after removing all duplicates. Ear from the profile face has been detected and cropped using a loose rectangular boundary by employing the technique proposed in [28]. Sample 3D ear images with their corresponding co-registered 2D ear images for four subjects of the database are shown in Fig. 5.3. In the experiment, one image of each subject has been considered as gallery image while remaining images are used as probe images. Flow diagram of LFGPA-ICP is shown in Fig. 5.4.

5.4.1 Preprocessing of Data

3D ear data sometimes exhibits noise in the form of spikes and holes caused due to oily skin or sensor error. In LFGPA-ICP technique, a preprocessing step is used to remove all such noise from the data. We apply median filtering to remove spikes in the data whereas linear interpolation is used to fill holes. Spikes in the data are removed by performing median filtering in 3×3 neighborhood. Filling is done only for the holes which are of size one pixel and have four valid neighbors.

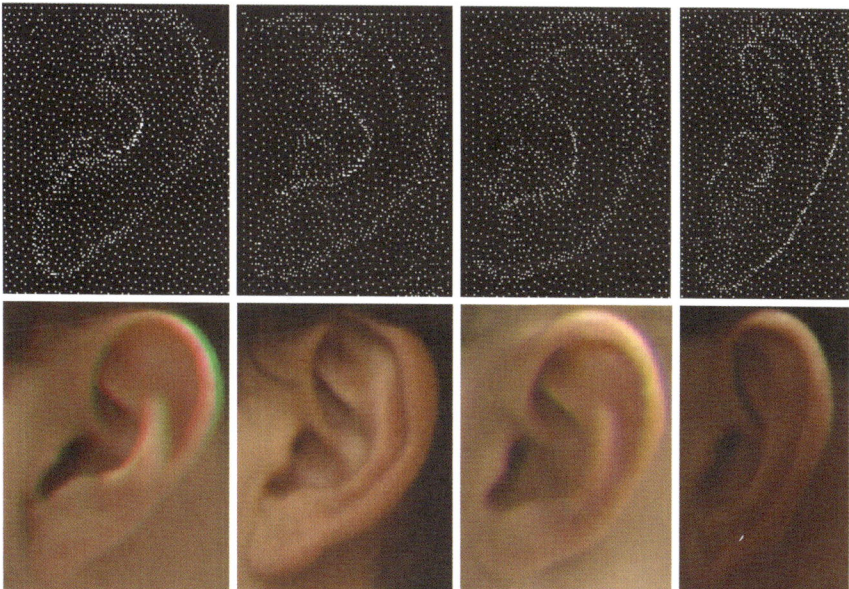

Fig. 5.3 3D ear images with their co-registered 2D ear images for four subjects (respective columns)

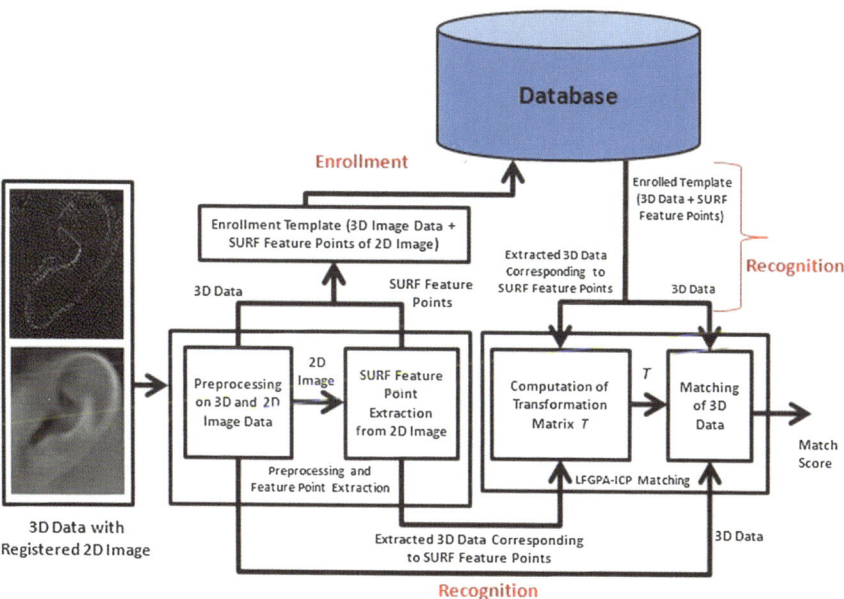

Fig. 5.4 Flow diagram of ear recognition technique (LFGPA-ICP)

5.4.2 Results

Ear recognition is performed using LFGPA-ICP technique where matching score is computed in terms of registration error obtained between the two 3D ear images being matched. A low registration error shows a good match whereas a high registration error reflects the poor match.

Experiments have been conducted by considering randomly selected one ear image of each subject as a gallery image and all other ear images of the subject as probe. Table 5.1 shows the recognition accuracy (with corresponding *FAR* and *FRR*), *EER* and *EUC* values for LFGPA-ICP. It has produced 98.30 % recognition accuracy with *FAR* as 1.2 % and *FRR* as 2.2 %. The values of *EER* and *EUC* are found to be 1.8 and 0.3 % respectively. *FAR* and *FRR* curves with respect to matching threshold are shown in Fig. 5.5a whereas *ROC* curve for the technique is shown in Fig. 5.5b.

5.4.3 Comparison with Other Techniques

Performance of LFGPA-ICP is analyzed with respect to other well known techniques in the literature. The technique proposed in [6] has carried out experiments on UND Collection F (UND-F) [29] databases (which is a subset of UND-J2 database) which consists of 942 ear images collected from 302 subjects. It has used 604 images of 302 subjects in the experiment and has achieved ear verification performance in terms of *EER* as 2.3 %. Another technique which has been proposed in [5] uses UND-J2 database for experiments and has achieved *EER* of 1.2 %. The technique proposed for identification in [11] has considered 200 ear images of the UND-F database and has reported rank-1, rank-2 and rank-3 recognition rates of 90, 94 and 96 % respectively. However, it has not been tested for verification. The technique in [12] has achieved 94.0 % accuracy with 4.1 % *EER* on 830 ear images of UND-J2 database.

LFGPA-ICP technique performs better than the techniques available in the literature except the one presented in [5] where a lower *EER* value is reported. However, the technique in [5] has used selective training where a good ear sample of a subject is considered for the gallery. Also, it has used a very concisely localized ear for matching. It has made use of Active Contours [30], 2D intensity and 3D curvature information for ear localization which has made the ear detection process computationally costly. Moreover, for concise ear localization, this technique relies

Table 5.1 Recognition results for LFGPA-ICP

Database	# of images	Accuracy (*FRR*, *FAR*) (%)	*EER* (%)	*EUC* (%)
UND-J2	1780	98.30 (1.2, 2.2)	1.8	0.3

Fig. 5.5 *FAR*, *FRR* and *ROC* curves of the LFGPA-ICP ear recognition technique. **a** Threshold versus FAR and FRR curves. **b** ROC curve

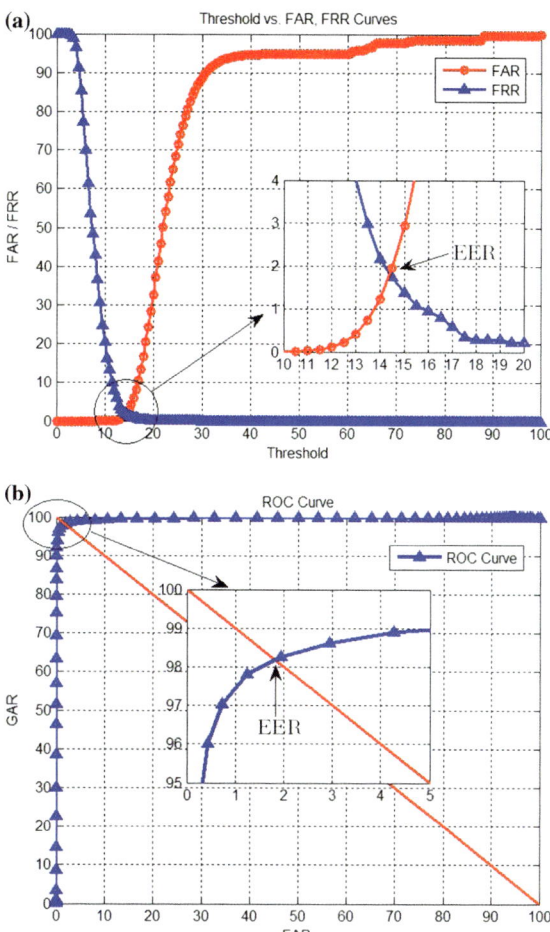

on estimation of nose tip and ear pit boundary. However, the visibility of nose tip and ear pit is very much sensitive to pose variations and hence, it is bound to fail when profile face deviates from the frontal straight-on position.

The type of evaluation and the experimental data used in LFGPA-ICP is similar to the one used in [6, 12]. The techniques presented in [6, 12] have also been evaluated for ear verification. These techniques have been tested on UND-J2 database by considering full or partial data. For example, experimental study in [6] has used UND-F [29] database which is a subset of the UND-J2 database. In [12], part of UND-J2 database has been used in experiment. Since experimental evaluation of the techniques presented in [6, 12] is similar to the one used in the LFGPA-ICP, we have presented a detailed performance comparison of LFGPA-ICP with these techniques in Table 5.2. It is evident from the table that LFGPA-ICP performs much better than both of these techniques. It is important to note that both these techniques

Table 5.2 Performance comparison with the state-of-the-art techniques

Technique	Database	Images used (gallery, probe)	Verification accuracy (%)	EER (%)
[6]	UND-F	604 (302, 302)	–	2.3
[12]	UND-J2	830 (415, 415)	94.00	4.1
LFGPA-ICP	**UND-J2**	1780 (404, 1376)	**98.30**	**1.8**

have used a subset of UND-J2 database for experimental evaluation. However on contrary, experimental evaluation of LFGPA-ICP has been done using full UND-J2 database which contains much large and diverse set of images. In spite of using large and diverse database, LFGPA-ICP has performed much superior to [6, 12] which not only shows its robustness over these techniques but also proves its scalability.

LFGPA-ICP performs superior to existing 3D ear recognition techniques due to following reasons. It has used a two-step 3D ear matching strategy. In the first step, it coarsely aligns 3D ear images using salient 3D data points obtained from these images with the help of local 2D feature points of co-registered 2D ear images. As GPA-ICP based matching is prone to get stuck into a local minima while convergence, the use of initial level coarse alignment of 3D ear images helps to avoid such situation. Also, initial coarse alignment helps in getting a good starting point for the second step (final) matching and in achieving fast and accurate final data alignment. In LFGPA-ICP, salient data points of a 3D ear image are computed with the help of local 2D feature points obtained from its co-registered 2D ear image, instead of directly computing them from 3D ear data. Since the field of local feature point extraction in 2D is much more matured as compared to that in 3D, computation of salient 3D data points with the help co-registered 2D ear image provides robust salient data points.

Another reason for superior performance of LFGPA-ICP is due to the integration of GPA with ICP. Such integrated technique (GPA-ICP) is more robust as compared to the traditional GPA due to the fact that it defines point correspondences between the two sets of data points being aligned by finding the mutual nearest neighbors which helps in defining true point correspondences. Also, GPA-ICP partitions the points of the data sets which are being aligned into independent sets (each independent set containing mutual nearest neighbor points) and computes centroid of each independent set using the points participating in them. Further, it performs the alignment of the data points of the two sets (which are being matched) with respect to the centroid points (rather than directly aligning the points of the two sets with each other) which provides robust matching and it extremely helps in reducing the alignment (registration) error.

Though LFGPA-ICP has been tested only on one database which is UND-J2 (the only database publicly available with 3D ear images along with co-registered 2D ear images), results obtained by LFGPA-ICP are generic and stable due to following reasons. First, they have been obtained on a fairly large sample size of UND-J2 database which consists of challenging ear images with pose and scale variations. Secondly, in experiments gallery image of a subject is selected randomly from the database in contrary to choosing a good image of the subject as done in [5]. Thirdly, results are averaged over multiple cycles by randomly selecting different sets of gallery and probe images.

References

1. Hui, Chen, and Bir, Bhanu. 2005. Contour matching for 3D ear recognition. In *Proceedings of IEEE Workshop on Application of Computer Vision (WACV/MOTION'05)*, vol. 1, 123–128.
2. Yan, Ping, and Kevin W. Bowyer. 2005. Ear biometrics using 2D and 3D images. In *Proceedings of International Conference on Computer Vision and Pattern Recognition-Workshop*, 121–128.
3. Yan, Ping, and Kevin W. Bowyer. 2005. Multi-biometrics 2D and 3D ear recognition. In *Proceedings of International Conference on Audio-and Video-Based Biometric Person Authentication*. LNCS, vol. 3546, 503–512.
4. Chen, H., B. Bhanu, and R. Wang. 2005. Performance evaluation and prediction for 3D ear recognition. In *Proceedings of International Conference on Audio and Video Based Biometric Person Authentication (AVBPA'05)*. LNCS, vol. 3546, 748.
5. Yan, Ping, and K.W. Bowyer. 2007. Biometric recognition using 3D ear shape. *IEEE Transactions on Pattern Analysis and Machine Intelligence* 29(8): 1297–1308.
6. Chen, Hui, and Bhanu, Bir. 2007. Human ear recognition in 3D. *IEEE Transactions on Pattern Analysis and Machine Intelligence* 29(4): 718–737.
7. Passalis, G., I.A. Kakadiaris, T. Theoharis, G. Toderici, and T. Papaioannou. 2007. Towards fast 3D ear recognition for real-life biometric applications. In *Proceedings of IEEE Conference on Advanced Video and Signal Based Surveillance (AVSS'07)*, vol. 3, 39–44.
8. Kirkpatrick, S., C.D. Gelatt Jr. and M.P. Vecchi. 1983. Optimization by simulated annealing. *Science* 220(4598): 671–680.
9. Cadavid, S., and M. Abdel-Mottaleb. 2007. Human identification based on 3D ear models. In *Proceedings of International Conference on Biometrics: Theory, Applications and Systems (BTAS'07)*, 1–6.
10. Islam, S.M.S., M. Bennamoun, A.S. Mian, and R. Davies. 2008. A fully automatic approach for human recognition from profile images using 2D and 3D ear data. In *Proceedings of 4th International Symposium on 3D Data Processing, Visualization and Transmission (3DPVT'08)*, 131–135.
11. Islam, S.M., R. Davies, A.S. Mian, and M. Bennamoun. 2008. A fast and fully automatic ear recognition approach based on 3D local surface features. In *Proceedings of 10th International Conference on Advanced Concepts for Intelligent Vision Systems (ACIVS'08)*, 1081–1092.
12. Islam, S.M.S., Rowan Davies, Mohammed Bennamoun, and Ajmal S. Mian. 2011. Efficient detection and recognition of 3D ears. *International Journal of Computer Vision* 95(1): 52–73.
13. Theoharis, Theoharis, Georgios Passalis, George Toderici, and Ioannis A. Kakadiaris. 2008. Unified 3D face and ear recognition using wavelets on geometry images. *Pattern Recognition* 41(3): 796–804.
14. Islam, S.M.S., Mohammed Bennamoun, Ajmal S. Mian, and R. Davies. 2009. Score level fusion of ear and face local 3D features for fast and expression-invariant human recognition. In *Proceedings of 6th International Conference on Image Analysis and Recognition (ICIAR'09)*, 387–396.
15. John, Bustard, and Mark, Nixon. 2010. 3D morphable model construction for robust ear and face recognition. In *Proceedings of International Conference on Computer Vision and Pattern Recognition (CVPR'10)*, 2582–2589.

16. Jindan, Zhou, S. Cadavid, and M. Abdel-Mottaleb. 2011. A computationally efficient approach to 3D ear recognition employing local and holistic features. In *Proceedings of IEEE Conference on Computer Vision and Pattern Recognition Workshop (CVPRW'11)*, 98–105.
17. Gower, J.C. 1975. Generalized procrustes analysis. *Psychometrika* 40(1): 33–51.
18. Schonemann, P. 1966. A generalized solution of the orthogonal procrustes problem. *Psychometrika* 31(1): 110.
19. Schonemann, P., and R. Carroll. 1970. Fitting one matrix to another under choice of a central dilation and a rigid motion. *Psychometrika* 35(2): 245–255.
20. Borg, Ingwer, and Patrick Groenen. 2005. *Modern multidimensional scaling: Theory and applications*. New York: Springer.
21. Commandeur, J.J.F. 1991. *Matching configurations*. Leiden University, Leiden, Netherlands: DSWO Press.
22. Roberto, Toldo, Alberto, Beinat, and Fabio, Crosilla. 2010. Global registration of multiple point clouds embedding the generalized procrustes analysis into an ICP framework. In *Proceedings of 5th International Symposium on 3D Data Processing, Visualization and Transmission (3DPVT'10)*.
23. Crosilla, F., and A. Beinat. 2002. Use of generalised procrustes analysis for the photogrammetric block adjustment by independent models. *ISPRS Journal of Photogrammetry and Remote Sensing* 56(3): 195–209.
24. Herbert, Bay, Ess Andreas, Tuytelaars Tinne, and Van Gool Luc. 2008. Speeded-up robust features (SURF). *Computer Vision and Image Understanding* 110(3): 346–359.
25. Bustard, J.D., and M.S. Nixon. 2008. Robust 2D ear registration and recognition based on SIFT point matching. In *Proceedings of International Conference on Biometrics: Theory, Applications and Systems (BTAS'08)*, 1–6.
26. Lowe, David G. 2004. Distinctive image features from scale-invariant keypoints. *International Journal of Computer Vision* 60(2): 91–110.
27. Prakash, Surya, and Phalguni Gupta. 2012. An efficient ear localization technique. *Image and Vision Computing* 30(1): 38–50.
28. Surya, Prakash, and Phalguni, Gupta. 2012. An efficient technique for ear detection in 3D: Invariant to rotation and scale. In *Proceedings of IAPR/IEEE International Conference on Biometrics (ICB'12)*, 97–102.
29. Yan, Ping, and Kevin W. Bowyer. 2005. Empirical evaluation of advanced ear biometrics. In *Proceedings of International Conference on Computer Vision and Pattern Recognition-Workshop*, vol. 3, 41–48.
30. Kass, M., A. Witkin, and D. Terzopoulos. 1988. Snakes: active contour models. *International Journal of Computer Vision* 1(4): 321–331.

Chapter 6
Fusion of Ear with Other Traits

6.1 Fusion in 2D

Fusion of ear with face is a good choice because of their physiological structure and location. Ear and face can be captured simultaneously using the same camera. Also, both of them can be acquired non-intrusively. Moreover, use of ear along with face in a biometric authentication system has a possibility to improve the accuracy and the robustness of the system, particularly for non-frontal views. Fusion of ear with face also seems to be more relevant from the perspective of surveillance application. To exploit these advantages, there exist multimodal techniques [1–3] which makes use of ear and face. These approaches have also extended to include other biometric modalities such as speech and gait.

In [1], appearance based features are used to fuse ear and face. It has used PCA to extract features from both the modalities and has shown that multimodal recognition using both ear and face has resulted in significant improvement over the unimodal traits. It has achieved a recognition rate of about 91 % in a database consisting of 88 probe and 197 gallery images. Further, Rahman and Ishikawa [3] have also proposed a multimodal biometric system using PCA on both face and ear. It has reported an improved recognition rate of 94.4 % in a multimodal database consisting of 90 ear and face images collected from 18 individuals in 5 sessions.

There exists multimodal system designed at Indian Institute of Technology Kanpur (IITK). To test the system, it makes use of a database containing ear, face, iris, palmprint and slap fingerprint images, at least two images of each individual per trait (one for gallery and one for probe). The decision is made by performing fusion at "matching score level". Feature vectors of each individual are compared with the enrolled templates (obtained for each gallery image) stored in the database for each biometric trait. Based on the proximity of the feature vector and the template, a matching score is computed for each trait. To analyze the performance of fusion of ear with other traits, individual scores of ear and face, iris, fingerprint and palmprint are combined to get a fused score which is used to make the authentication decision. Table 6.1 presents the detailed information of database of various traits.

© Springer Science+Business Media Singapore 2015
S. Prakash and P. Gupta, *Ear Biometrics in 2D and 3D*,
Augmented Vision and Reality 10, DOI 10.1007/978-981-287-375-0_6

Table 6.1 Unimodal database

Database	Distinct subjects/types
Ear	279
Face	307
Iris	1019
Palmprint	163
Slap fingerprint	990

By distinct types in the table, it is intended to refer images of left or right hand/ear. These images are artificially combined to obtain a multimodal database. Sample images are shown in Fig. 6.1. Accuracy when only one trait is used for recognition is provided in Table 6.2. Distribution of genuine scores against imposter scores for

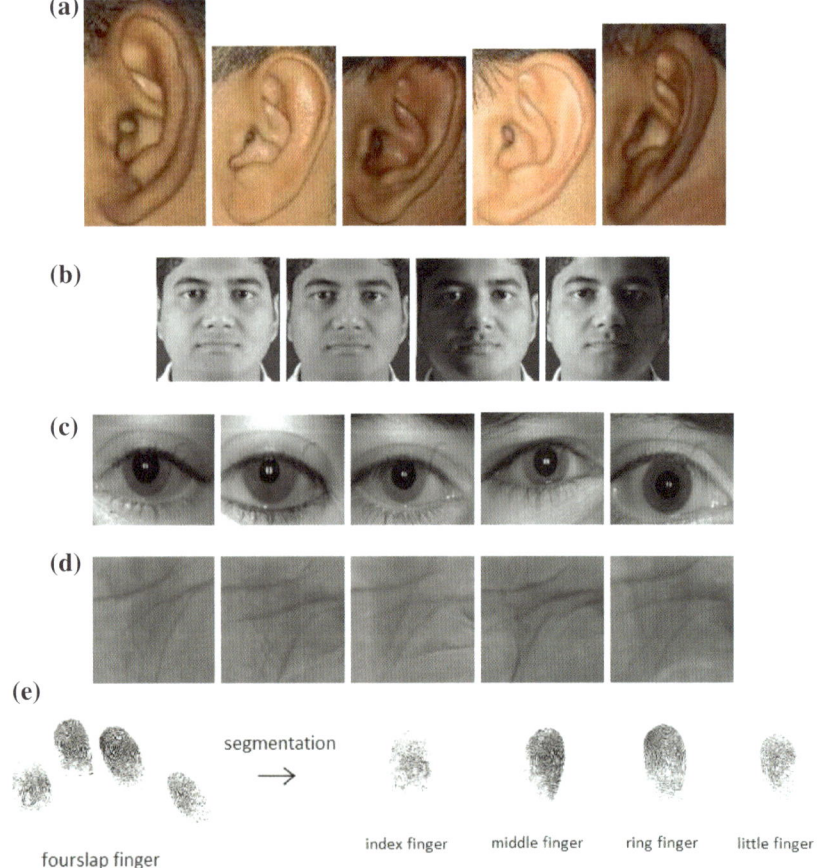

Fig. 6.1 Sample images from IITK database. **a** Ear images. **b** Face images. **c** Iris images. **d** Palmprint images. **e** Slap fingerprint image

Table 6.2 Recognition results when only one trait is used

Database	Subjects	CRR	EER	Accuracy
Ear	279	86.23	0.71	99.29
Face	307	98.02	0.97	99.03
Iris	1019	98.52	0.70	99.30
Palmprint	163	84.38	8.94	91.06
Slap fingerprint	990	99.69	0.60	99.40

each trait is shown on Fig. 6.2. It is observed that slap fingerprint system when all four fingers are used performs the best among the five traits. Performance of the biometric system is improved when fusion of ear with various other biometric traits is considered. Table 6.3 summarizes the recognition results. It can be seen that in all fusion experiments, 100 % Correct Recognition Rate (CRR) has been achieved.

1. **Fusion of Ear and Face**: The face database consists of 307 subjects while there are only 279 subjects available in ear database. Hence in multimodal experiment, a database of 279 subjects having both face and ear images is considered. It is observed that the accuracy of the system is 99.67 %. Distribution of genuine score against imposter score for this system shown in Fig. 6.3 shows their separability. ROC curve for the fused system is shown in Fig. 6.4.
2. **Fusion of Ear and Iris**: The iris database consists of 1019 subjects while ear database has images only from 279 subjects. The database used in this fusion experiment consists of only 279 subjects having both ear and iris images. The proposed fusion has achieved accuracy of 99.99 %. Figure 6.5 shows the genuine against imposter score distribution. Further, ROC curve for this fusion is shown in Fig. 6.6.
3. **Fusion of Ear and Fingerprint**: Using fingerprint and ear databases, a combined database of 279 subjects having both ear and fingerprint images has been created to test the multimodal system. It is observed that this fusion has achieved an accuracy of 100.00 %. Distribution of genuine against imposter scores is drawn in Fig. 6.7 while ROC curve for this experiment is shown in Fig. 6.8.
4. **Fusion of Ear and Palmprint**: Using ear database of 279 subjects and palmprint database of 163 subjects, a multimodal database of 163 subjects having both ear and palmprint images has been created. The multimodal system based on fusion of ear and palmprint has been tested on this database. It is observed that the fusion of ear and palmprint has achieved a verification accuracy of 99.37 %. Distribution of genuine against imposter scores and its ROC curve for this experiment are shown in Figs. 6.9 and 6.10 respectively.

Further, Iwano et al. [4] have integrated ear image with the speech information to improve the robustness of person authentication. A multimodal biometric person authentication technique has been proposed which makes use of speech and ear images to increase the performance in mobile environment. It has combined ear

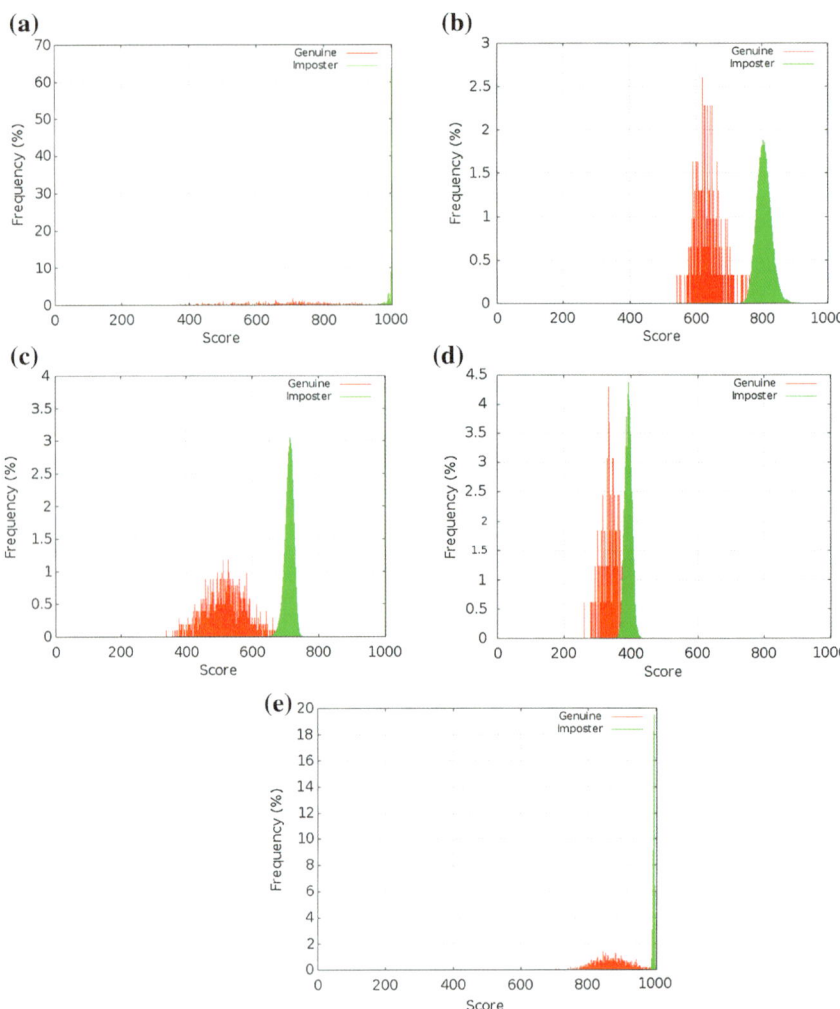

Fig. 6.2 Genuine versus imposter score histograms for various traits. **a** Ear. **b** Face. **c** Iris. **d** Palmprint. **e** Slap fingerprint

images with speech using a composite posterior probability. An audio-visual database collected from 38 male speakers in five sessions, acquiring one speech and one ear sample in each session, has been used to perform experiments. It has demonstrated the improvement over the system using alone either ear or speech. It has also shown that the improvement in authentication performance due to combining of ear image with speech happens in every SNR (signal-to-noise ratio) condition.

There exist multi-algorithmic approaches to achieve the improved authentication using ear. For example, Iwano et al. [2] have demonstrated a multi-algorithmic approach of ear recognition on audio-visual database (speech and ear images) where

Table 6.3 Recognition results when two traits are fused

Database	Subjects	CRR	EER	Accuracy
Ear, face	279	100	0.33	99.67
Ear, iris	279	100	0.01	99.99
Ear, palmprint	163	100	0.63	99.37
Ear, slap fingerprint	279	100	0.00	100.00

Fig. 6.3 Genuine versus imposter score histogram for fusion of ear and face

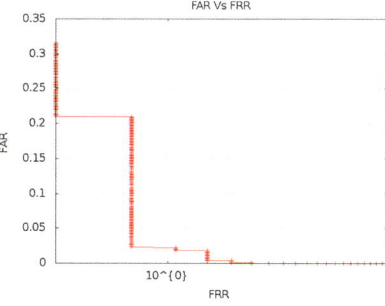

Fig. 6.4 ROC curve for fusion of ear and face

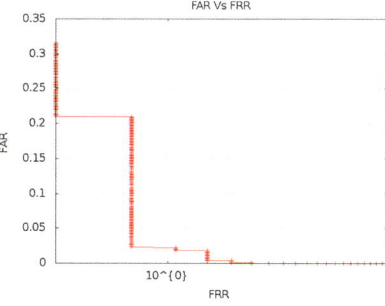

Fig. 6.5 Genuine versus imposter score histogram for fusion of ear and iris

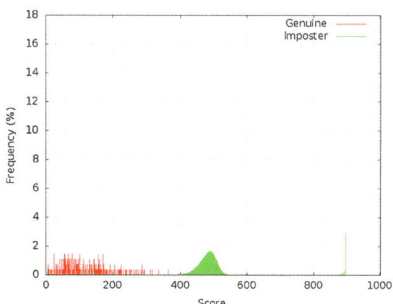

features from ear images are extracted using PCA and ICA. It compares performance of two authentication techniques using ear images, in which features are extracted by either PCA or ICA. Further, it has investigated the effectiveness of combining PCA

Fig. 6.6 ROC curve for
fusion of ear and iris

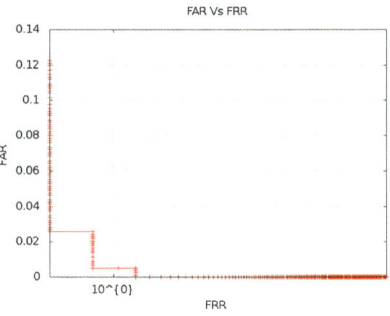

Fig. 6.7 Genuine versus
imposter score histogram for
fusion of ear and fingerprint

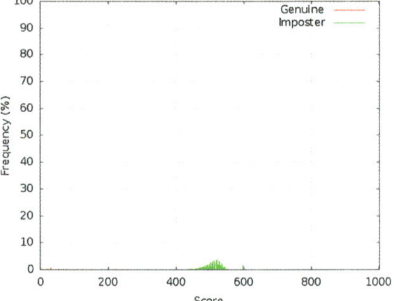

Fig. 6.8 ROC curve for
fusion of ear and fingerprint

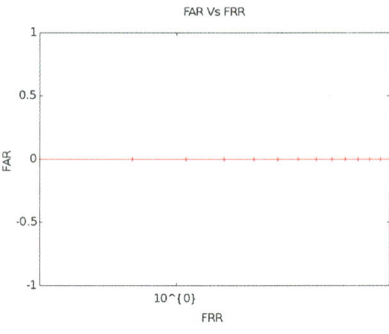

Fig. 6.9 Genuine versus
imposter score histogram for
fusion of ear and palmprint

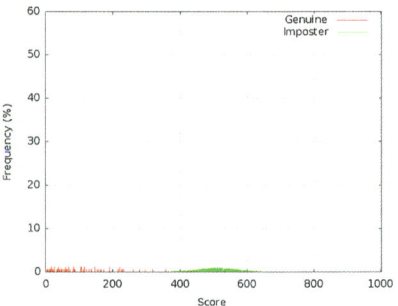

Fig. 6.10 ROC curve for fusion of ear and palmprint

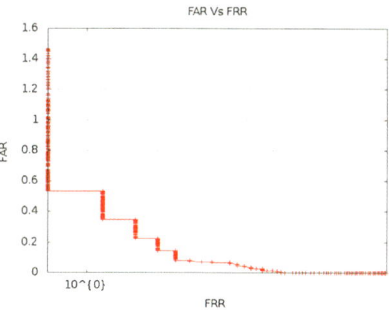

and ICA based features for authentication. It has shown improvement as compared to one when either of the feature extraction is applied alone. Since the fusion of PCA and ICA based ear authentication technique is found to be effective, it is concluded that the two authentication techniques are sufficiently independent. It has also been shown that combination of PCA and ICA features of ear along with speech information results into a stronger multimodal system.

6.2 Fusion in 3D

Fusion of ear with other traits has been considered in 3D as well. Since it has been established that there is a low correlation between the differentiability of 3D face and ear data [5], features of ear and face have improved the performance.

In [5], a unified technique which fuses 3D ear and facial data has been proposed. It uses an annotated deformable model of the ear and the face to extract respective geometry images from their respective data. Further, it computes wavelet coefficients from these geometry images. Wavelet coefficients are used as biometric signature for recognition. It has achieved the rank-1 accuracy of 99.7 % on a database of 648 pairs of 3D face and ear range images obtained from 324 subjects. The multimodal database has been created by considering face images of FRGC v2 database [6] and ear images of Ear Database (Collections F and G) of the University of Notre Dame (UND) [7].

In [8], an expression-robust multimodal ear and face based biometric technique has been proposed. It has considered the fusion at score level and is based on local 3D features which are fast to compute and are robust to pose and scale variations. These features are also robust to occlusion due to hair and earrings. To extract local 3D features from 3D ear and 3D face data, a number of distinctive 3D feature point locations (keypoints) are selected from 3D ear and 3D face regions using asymmetrical variation in depth around them. Asymmetrical variation in depth is computed through the difference between the first two eigenvalues in PCA of the data window centred on the keypoints [9]. It is observed that for different individuals, number and locations of the keypoints are different for the ear and the face images. It is also seen that the keypoints have high degree of repeatability for the same individual. Matching of the features in this technique is done at two levels. At coarser level, features are matched using Euclidean distance and rotation angles between the underlying coordinate bases of the features [10]. These angles are further clustered and the largest cluster is used for coarse alignment of probe and gallery images. For final alignment, Iterative Closest Point (ICP) matching algorithm [11] has been used. The technique has been tested on a multimodal database which includes 326 gallery images with 311 and 315 probe images with neutral and non-neutral expressions respectively. A multimodal database has been created by considering face images of FRGC v2 database [6] and ear images of Ear Database (Collection J2) of the University of Notre Dame (UND) [12]. It has achieved an identification accuracy of 98.71 % and a verification accuracy of 99.68 % for the fusion of the ear with the neutral face. Further, a recognition rate of 98.10 % with a verification accuracy of 96.83 % has been achieved when there exist expressions in facial images.

There have been multiple attempts in the field of face recognition to use morphable model fitting to recognize people under relatively unconstrained environment. The technique in [13] makes the efficient construction of 3D morphable model of the head and the ear for human recognition. This model of the head and the ear statistically builds a joint model based on 3D shape and 2D texture of the head and the ear respectively. These models are explicitly designed to recreate both the face and the ear shape accurately. Training datasets used to create morphological model include ear images with noise and partial occlusion due to hair. This imposes challenges in the construction of correct morphological model and often, these erroneous regions are manually excluded in the creation of the model. In this technique, rather than excluding these regions manually, a classifier has been developed to automate this process. The model is developed by registering a generic head mesh with multiple range scans of face and ear profiles. Regions having occlusion and noise are identified within each scan using the automated classifier. The remaining valid regions are used to register the mesh using a robust non-linear optimisation algorithm which enables efficient construction of full head morphable model using less constrained datasets. Orientations of the scans are normalized and are further used to construct a linear model of all head shapes.

A general 3D object recognition technique by combining local and holistic features has been proposed in [14]. The technique has been evaluated for 3D ear recognition task. It has primarily focused on local and holistic feature extraction and matching components, in addition to fusion framework to combine these features at the matching score level. For the local feature representation and matching, the technique has introduced Histogram of Indexed Shapes (HIS) feature descriptor and has extended it to Surface Patch Histogram of Indexed Shapes (SPHIS) which is an object centered 3D shape descriptor. For holistic feature extraction and matching, voxelization of the ear surface is carried out to generate a representation from which an efficient voxel-wise comparison of gallery-probe model pairs can be done. The matching scores obtained from local and holistic matching components are fused

Table 6.4 Recognition
results reported in [16]

Technique used	Recognition rate (%)
2D PCA	71.9
3D PCA	64.8
3D Edge	71.9
3D PCA + 3D Edge	80.2
2D PCA + 3D Edge	89.7
2D PCA + 3D PCA	89.1
2D PCA + 3D PCA + 3D Edge	90.6

to obtain the final match scores. The technique has yielded a rank-1 recognition accuracy of 98.6 % and an *EER* of 1.6 % on UND-G [15] database.

Yan and Bowyer [16] have carried out multimodal experiments to test the performance improvement for various combinations of 2D-PCA, 3D-PCA and 3D-Edges in ear recognition. Experimental results are shown in Table 6.4.

Though ear and face based multimodal techniques have achieved improved performance, they are computationally expensive for the large volume of 3D ear and the face data and hence seems to have low practical applicability. There is a need to develop fast representation and matching algorithms for 3D ear with other traits to make these systems practically applicable.

References

1. Chang, Kyong, Kevin W. Bowyer, Sudeep, Sarkar, and Barnabas, Victor. 2003. Comparison and combination of ear and face images in appearance-based biometrics. *IEEE Transactions on Pattern Analysis and Machine Intelligence* 25(9): 1160–1165.
2. Iwano, K., T. Miyazaki, and S. Furui. 2005. Multimodal speaker verification using ear image features extracted by PCA and ICA. *Proceedings of International Conference on Audio and Video based Biometric Person Authentication*. LNCS, vol. 3546, 588–5996.
3. Rahman, M.M., and Ishikawa. S. 2005. Proposing a passive biometric system for robotic vision. In *Proceedings of 10th International Symposium on Artificial Life and Robotics (AROB'05)*, 4–6.
4. Iwano, K., Hirose, T., Kamibayashi, E., and Furui. S. 2003. Audio-visual person authentication using speech and ear images. In *Proceedings of Workshop on Multimodal User Authentication*, 85–90.
5. Theoharis, Theoharis, Georgios, Passalis, George, Toderici, and Ioannis A. Kakadiaris. 2008. Unified 3D face and ear recognition using wavelets on geometry images. *Pattern Recognition* 41(3): 796–804.
6. Phillips, P.J., Flynn, P.J., Scruggs, T., Bowyer, K.W., Chang, J., Hoffman, K., Marques, J., Min, J., and Worek. W. 2005. Overview of the face recognition grand challenge. In *Proceedings of Computer Vision and Pattern Recognition, (CVPR 2005)*, 947–954.
7. University of Notre Dame Profile Face Database. Collections F and G. http://www.nd.edu/~cvrl/CVRL/DataSets.html.

8. Islam, S.M.S., Mohammed, Bennamoun, Ajmal, S., Mian, and Davies. R. 2009. score level fusion of ear and face local 3D features for fast and expression-invariant human recognition. In *Proceedings of 6th International Conference on Image Analysis and Recognition (ICIAR'09)*, 387–396.

9. Mian, Ajmal S., Mohammed, Bennamoun, and Robyn, Owens. 2008. Keypoint detection and local feature matching for textured 3d face recognition. *International Journal of Computer Vision* 79(1): 1–12.

10. Islam, S.M., Davies, R., Mian, A.S., and Bennamoun. M. 2008. A fast and fully automatic ear recognition approach based on 3D local surface features. In *Proceedings of 10th International Conference on Advanced Concepts for Intelligent Vision Systems (ACIVS'08)*, 1081–1092.

11. Mian, Ajmal, Mohammed, Bennamoun, and Robyn, Owens. 2007. An efficient multimodal 2d–3d hybrid approach to automatic face recognition. *IEEE Transactions on Pattern Analysis and Machine Intelligence* 29(11): 1927–1943.

12. University of Notre Dame Profile Face Database. Collection J2. http://www.nd.edu/~cvrl/CVRL/DataSets.html.

13. John, Bustard, and Mark, Nixon. 2010. 3D morphable model construction for robust ear and face recognition. In *Proceedings of International Conference on Computer Vision and Pattern Recognition (CVPR'10)*, 2582–2589.

14. Jindan, Zhou, Cadavid, S., and Abdel-Mottaleb, M. 2011. A computationally efficient approach to 3D ear recognition employing local and holistic features. In *Proceedings of IEEE Conference on Computer Vision and Pattern Recognition Workshop (CVPRW'11)*, 98–105.

15. University of Notre Dame Profile Face Database, Collection G. http://www.nd.edu/~cvrl/CVRL/DataSets.html.

16. Yan, P., and Bowyer, K.W. 2004. 2D and 3D ear recognition. In *Proceedings of Biometric Consortium Conference*.